Construction Planning, Programming and Control

Construction Planning, Programming and Control

Second Edition

Brian Cooke and Peter Williams

© 1998, 2004 Brian Cooke and Peter Williams

Blackwell Publishing Ltd,
Editorial offices:
Blackwell Publishing Ltd, 9600 Garsington Road, Oxford OX4 2DQ, UK
  Tel: +44 (0)1865 776868
Blackwell Publishing Inc., 350 Main Street, Malden, MA 02148-5020, USA
  Tel: +1 781 388 8250
Blackwell Publishing Asia Pty Ltd, 550 Swanston Street, Carlton, Victoria 3053, Australia
  Tel: +61 (0)3 8359 1011

First published 1998 by Palgrave Publishers Ltd
Second edition published 2004 by Blackwell Publishing

Library of Congress Cataloging-in-Publication Data

Cooke, B. (Brian)
  Construction planning, programming, and control / Brian Cooke and Peter Williams.— 2nd
ed.
    p. cm.
  Includes bibliographical references and index.
  ISBN 1–4051–2148–3 (pbk. : alk. paper)
  1. Building—Planning. I. Williams, Peter. II. Title.

  TH153.C597 2004
  690′.068—dc22

                                                                    2004003227

A catalogue record for this title is available from the British Library

Set in 10/12.5 pt Palatino
by Graphicraft Limited, Hong Kong
Printed and bound in the UK
by TJ International Ltd, Padstow, Cornwall

The publisher's policy is to use permanent paper from mills that operate a sustainable forestry policy,
and which has been manufactured from pulp processed using acid-free and elementary chlorine-free
practices. Furthermore, the publisher ensures that the text paper and cover board used have met
acceptable environmental accreditation standards.

For further information on Blackwell Publishing, visit our website:
www.thatconstructionsite.com

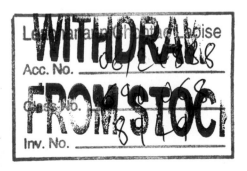

Contents

Preface

The aims of the first edition have not changed. This is a book for students of construction related professional and degree courses – construction management, building, quantity surveying and building surveying alike. Students of all these topics need a basic understanding of planning, procurement and programming at some time in their career development.

The text has been extensively rewritten and restructured. This is to reflect the extent of changes in the construction industry due particularly to the Latham and Egan Reports and recommendations from *Modernising Construction* (National Audit Office).

An overview has been included of the various procurement options available to the client; the advantages and disadvantages of traditional, design and build alternatives, construction management, partnering and PFI have all been considered from a practical angle. Framework and prime contracting have also been considered.

The overall planning process has been related to both the client and contractor. Pre-tender planning procedures have been explained both on traditional contracts and design and build arrangements.

Project control has been considered from the angle of time, money and resources control. The control of money has been extended to include the principles and application of cash flow and considerations for improving the cash flow in a business.

The programming techniques available to the contractor do not change, just the ways in which they are applied to a project. However, the use of procurement programmes and information schedules has moved to the fore as more emphasis is placed on design and build and partnering, and this has been reflected in the examples and case studies in this book.

Our special thanks go to Julia Burden of Blackwell Publishing for supporting the second edition of this publication. Paul Hodgkinson has again ably assisted us with the diagrams.

Finally, the book could not have been written without the support of the directors and staff of the Totty Construction Group who have allowed us access to their company procedures and the City Road Project.

1 Procurement and contracts

1.1 The construction industry

The effective planning and control of construction projects requires the application of systematic and logical methods and tried and tested techniques aimed at ensuring successful project outcomes for the client, the contractor and all other project participants. Control is not possible without a plan, and without a programme there is no effective means of exercising control.

Successful projects cannot happen in a vacuum, however, and no-one can effectively plan and control a construction project without understanding the culture and methodologies of the industry that organises and carries out the work, and the impact of the various procurement strategies open to the client.

Construction is a large and complex industry comprising many types and sizes of organisations and a diverse range of professional and other representative bodies. Clients, professional practices, contractors and specialist firms all have their own 'agenda' and allegiances; Walker (2002) explains the complexities of the interrelationships that can occur in construction projects between clients and their professional advisors and the contractors and specialists they engage to construct the project.

The annual output of the UK construction industry is approximately £65 billion which represents about 8% of GDP. The industry is also a large employer with around 1.9 million people, of which 0.5 million are self-employed. Construction has a unique structure with a small number of very large firms carrying out the majority of its turnover. It also has a very large number of relatively small firms with few criteria limiting entry to the industry.

There is a wide disparity in the standards of competence across the industry, a largely semi-skilled and itinerant work force and a generally low standard of education and qualifications in the managers employed in construction.

The National Audit Office (2001) reports that the industry is made up of some 163 000 firms of which:

- 1% are large firms employing over 80 people
- 4% are medium-sized employing between 14 and 79 people
- 95% are small firms employing less than 14 people

The large contractors have considerable influence in the industry as they enjoy a large share of the construction market. Consequently, the Major Contractors Group (MCG) have lined up behind the Construction Skills Certification Scheme (CSCS) in order to help promote greater competence and attention to health and safety risks. CSCS is a voluntary registration scheme whereby trainees, workers, managers, supervisors, visiting professionals, etc. can obtain the relevant colour card which enables them to gain access to MCG sites. A health and safety awareness test must be passed in order to obtain the card.

The industry has a diverse range of suppliers as well as contractors, including manufacturers of materials and components, suppliers of quarry products and ready-mixed concrete, builders' merchants and plant hire firms. Some of the industry's suppliers are larger than the largest contractors working in the industry and they also have considerable influence on the way the industry operates.

One of the big problems in construction is the extent to which the industry separates design from production to a far greater extent than in other industries. This particular feature of the industry is still common despite the deficiencies of traditional procurement and the benefits offered by newer and more flexible approaches.

The industry is a highly fragmented project-based industry and leadership in construction comes from clients and not from the contractors and specialists who carry out the work. In 1998, the Construction Clients' Forum (now the Confederation of Construction Clients) established the Clients' Charter which sets out the minimum standards in procurement expected by clients. By registering for the Charter, clients commit themselves to establishing a modern business culture with their suppliers, steady improvement of standards measured against nationally accepted criteria and the exchange of best practice experience.

Consequently it is the clients and their professional advisors who dictate the procurement methods used in the industry and it is the contractors who have to react to the latest 'flavour of the month' in the context of the organisation and management of the project.

The scale and diversity of the clients, contractors, professionals and others who are involved in construction is enormous. A contractor may be a small 'sole trader' or a large public company and a client could be a domestic householder, private sector corporation or government department.

This means that there is nothing 'typical' about the customs and practices in construction. More or less every project has its own individuality and peculiarities depending upon the site and location, the design and type of construction, the business arrangements between the parties and the hopes and expectations of all those involved.

There is, however, one common feature and that is the 'project-based' nature of the industry. The whole focus is on the project; the time, cost, quality, resources, problems and solutions are all geared to the project.

1.2 Construction industry reports

In order to stimulate debate about construction industry practices and procedures, a number of well-known reports have been published over the years, including those shown in Table 1.1.

Several of the reports published prior to Latham and Egan raised similar criticisms about the customs and practices of the industry. The Emmerson Report, for instance, emphasised the need to improve co-ordination between participants in the construction process, while Banwell tabled concerns about the lack of early contractor involvement and the need for greater integration of subcontractors in the building team, and he even suggested that a common form of contract should be adopted for use on all construction projects.

These problems, and many others, have been recognised since World War II but despite identifying the problems and proposing solutions, most of the reports have had little influence on either government or the industry over the years.

Table 1.1 Construction industry reports.

Report	Title	Year
Simon Report	The Placing and Management of Building Contracts	1944
Emmerson Report	Survey of Problems Before the Construction Industries	1962
Banwell Report	The Placing and Management of Contracts for Building and Civil Engineering Work	1964
National Economic Development Office	Action on Banwell	1967
Tavistock Report	Interdependence and Uncertainty	1966
Latham 1	Interim Report – Trust and Money	1993
Latham 2	Final Report – Constructing the Team	1994
Levene Efficiency Scrutiny	Construction Procurement by Government	1995
Egan Report 1	Rethinking Construction	1998
National Audit Office	Modernising Construction	2001
Egan Report 2	Accelerating Change	2002

Latham – 'Trust and Money'

Perhaps the most influential of all the reports concerning the industry and its problems is 'Constructing the Team' written by Sir Michael Latham (1994) who was commissioned by both Government and the industry to 'Review the Procurement and Contractual Arrangements in the UK Construction Industry'.

Prior to final publication of his report in July 1994, Sir Michael produced an interim report in December 1993 called 'Trust and Money'. This report raised concerns about the extent of mistrust between professionals and contractors and contractors and subcontractors in construction, and also flagged up the endemic culture of late and conditional payments operating in the industry. The prevailing atmosphere of mistrust and slow payments was reported to result in disharmony in project teams, poor standards of work and poor client satisfaction.

Latham – 'Constructing the Team'

'Constructing the Team' is better known as the 'Latham Report' and its purpose was to find ways to 'reduce conflict and litigation and encourage the industry's productivity and competitiveness'.

The specific terms of reference for the review were to consider:

- Current procurement and contractual arrangements
- Current roles, responsibilities and performance of the participants, including the client

The report took account of the structure of the industry and the need for fairness, accountability, quality and efficiency and paid particular regard to:

- Client briefing
- Procurement methods
- The design process
- The construction process
- Contractual issues
- Dispute resolution

The report runs to some 130 pages and contains 30 main observations and recommendations, with the principal emphasis being on 'teamwork' in order to achieve 'win-win' solutions.

Latham noted several issues which influence the ability of the construction industry to respond effectively to its customers' requirements, and these can be summarised briefly as:

- Sensitivity to changes in government spending patterns
- Intense competition for work
- Inability to respond to increased demand
- Lack of competency testing of firms/workers entering the industry
- Lack of training
- Mistrust between the participants in construction projects
- Inadequate capital base (i.e. most contractors are undercapitalised)
- Adversarial attitudes
- Claims-conscious contractors
- High levels of insolvency

Some of the main points made by Latham clearly have important consequences for the planning, production and control of construction and are therefore directly relevant to this book. These include:

- The need for a set of basic principles for modern contracts
- Greater use of the New Engineering Contract which could become a common contract for the whole industry
- Improved tendering arrangements and more advice on partnering arrangements
- Evaluation of tenders on quality as well as price
- Fairer treatment of subcontractors with particular regard to tendering and teamwork on site
- A real cost reduction target in construction of 30% by the year 2000
- Pay when paid contract terms to be outlawed
- Adjudication to be the normal method of dispute resolution
- Fair contract terms backed up by a legislation
- Insolvency protection by means of trust funds

One of the key issues considered by Latham was the productivity of the industry and Latham clearly considered that this is linked to the quality of design preparation and information. Inefficiency creeps in where designs are incomplete or information given to the contractor is conflicting or too late to allow proper planning of production.

An issue of major importance is conflict in the industry both between clients and contractors and between contractors and their subcontractors. Latham suggested that considerable efficiencies can be gained by making changes in 'procurement practice, contract conditions, tighter restrictions over set-off and the introduction of adjudicators as a normal procedure for settling disputes'. He also concluded that the 'most effective form of contract in modern conditions should include:

- A specific duty for all parties to deal fairly with each other, and with their subcontractors, specialists and suppliers, in an atmosphere of mutual cooperation

- Taking all reasonable steps to avoid changes to pre-planned works information. But, where variations do occur, they should be priced in advance, with provision for independent adjudication if agreement cannot be reached'

and that 'subcontractors should undertake that, in the spirit of teamwork, they will coordinate their activities effectively with each other, and thereby assist the achievement of the main contractor's overall programme. They may need to price for such interface work.'

The conclusions of the review were clearly extensive and led to the formation of the Construction Industry Board (subsequently replaced by the Strategic Forum) and an extensive programme of initiatives including:

- Steps to improve productivity
- Better quality design and improved briefing of designers
- Changes to trade and professional training and education
- Improving the image of the industry
- Encouraging fewer disputes by encouraging partnering between contracting parties
- Improved quality of construction professionals, contractors and subcontractors
- The publication of reports from 12 working groups concerning subjects as diverse as client briefing, education and partnering

Some of the Latham recommendations were included in the Housing Grants, Construction and Regeneration Act 1996.

Egan – 'Rethinking Construction'

'Rethinking Construction' was published in July 1998 (Construction Task Force 1998) and represents the work of a special task force which was set up by the Government to identify the scope for improving quality and efficiency in construction. The task force was chaired by Sir John Egan, hence the popular title for the report – the Egan Report.

The Egan Report at 40 pages is certainly not as comprehensive as its predecessor, the Latham Report, but is no less searching and probably considerably more controversial. It contains many 'home truths' but may also be said to contain unfair criticisms, particularly with respect to comparisons with factory-based manufacturing industries such as the motor industry.

Latham looked at designing an infrastructure for the industry aimed at removing the inefficiencies and inconsistencies, especially in terms of client briefing, better design management and more coherent project strategies. In *Rethinking Construction* there is no industry 'blueprint' for change but the Construction Task Force, which produced the report, took the lead on a number of new initiatives, including:

- Movement for Innovation (known as m4i) – a board of members whose task is to coordinate a number of demonstration projects, to disseminate best practice information and to oversee industry-wide benchmarking.
- The Construction Best Practice Programme provides information for firms wanting to improve their performance.
- The idea behind Inside UK Enterprise was for top performing companies to have an 'open day' where other firms could visit and find out how things are done by the 'host' company.

The Egan Report has probably had more publicity than the Latham Report and certainly there has been plenty of action as a result of the report, including the introduction of Key Performance Indicators (KPIs) and Demonstration Projects exemplifying best practice.

The Egan Report undoubtedly recognises both the good and bad in construction and seeks to build on those aspects of the industry which are excellent in a worldwide context. However, on balance, the conclusion of the report is that the industry as a whole is underachieving and that there should be radical change in key areas of its performance. These include quality, productivity, cost and time certainty and health and safety.

In the Executive Summary, the Egan Report makes the following observations:

- The UK construction industry at its best is excellent. Its capability to deliver the most difficult and innovative projects matches that of any other construction industry in the world.
- Nonetheless, there is deep concern that the industry as a whole is underachieving. It has low profitability and invests too little in capital, research and development and training. Too many of the industry's clients are dissatisfied with its overall performance.
- If the industry is to achieve its full potential, substantial changes in its culture and structure are also required to support improvement. The industry must provide 'decent and safe working conditions and improve management and supervisory skills' at all levels. The industry must design projects for ease of construction making maximum use of standard components and processes.
- The industry must replace competitive tendering with 'long term relationships based on clear measurement of performance and sustained improvements in quality and efficiency'.

The Egan Report identified five key drivers of change needed to set the agenda for the industry:

(1)   Committed leadership
(2)   A focus on the customer
(3)   Integrated processes and teams
(4)   A quality driven agenda
(5)   Commitment to people

Among the year-on-year targets proposed by Egan were:

- 10% reduction in construction time from client approval to practical completion
- 10% increase in productivity
- 20% reduction in the number of reportable accidents
- 10% increase in turnover and profits of construction firms

One of the problems with the Egan Report is that the emphasis is placed on the 'top-end' of the industry, whereas Latham looked at the fundamental problems of the entire industry. So while Egan has led to the development of several good ideas and worthwhile aims, the concepts may take some time to filter down to the lower echelons of the industry.

Egan – 'Accelerating Change'

This report presents the first year's work of the Strategic Forum for Construction which replaced the now defunct Construction Industry Board that was set up following the Latham Report.

'Accelerating Change' identifies ways of increasing the pace of change following the recommendations in 'Rethinking Construction', reports on the progress made to date and sets out a strategic direction with targets. The report identifies three main drivers to accelerate change in construction and introduce a culture of continuous improvement in the industry:

- The need for client leadership
- The need for integrated teams and supply chains
- The need to address 'people issues', especially health and safety

The vision and aspirations set out in 'Accelerating Change' emphasise the need for collaboration between the whole supply team, including clients and manufacturers. The report represents a 'manifesto for change' for all involved in construction, including government, schools and further/higher education and professional bodies.

The strategic targets identified in 'Accelerating Change' include:

- By the end of 2004
  - 20% of construction projects by value to be undertaken by integrated teams and supply chains
  - 20% of clients to adopt the principles of the Client's Charter
  - 10% annual improvement by adopting the Client's Charter
- By the end of 2007
  - These figures rising to 50%
- By the end of 2006
  - 300 000 qualified people to be recruited to the industry
- By 2007
  - 50% increase in applications for built environment courses
- No later than 2010
  - A certificated fully trained, qualified and competent workforce

The report suggests six keys steps that clients must consider when considering whether or not to build:

(1) Verification of need
(2) Assessment of options
(3) Develop procurement strategy
(4) Implement procurement strategy
(5) Project delivery
(6) Post project review

National Audit Office – 'Modernising Construction'

This 103 page report by the National Audit Office (2001) recognises that there are inefficiencies in the traditional methods of procuring and managing major projects and, especially, in the practice of awarding contracts on the basis of lowest price. The report suggests that this does not provide value for money either in terms of initial costs or through-life and operational costs.

According to the report, 73% of construction projects carried out by government departments and agencies were over budget and 70% were delivered late. It is also reported that a benchmarking study of 66 central government construction projects (total value of £500 million) carried out in 1999 showed that three-quarters of projects exceeded their budgets by up to 50% and two-thirds exceeded their original completion date by 63%.

The report suggests that a major contributory factor to this poor performance is the adversarial relationships that exist between construction firms, consultants and clients and between contractors, subcontractors and suppliers. It is further suggested that the entire supply chain must be integrated and that risk and value must be managed in order to improve buildability, reduce accidents and drive out waste.

'Modernising Construction' identifies that government departments and agencies have adopted three main strategies for ensuring more collaboration, integration and value for money in the procurement of construction services, including:

- PFI (projects over £20 million)
- Prime contracting
- Design and build

The report emphasises the benefits of long-term collaborative relationships between clients and contractors (partnering). It is emphasised that partners should be appointed competitively and that a hands-off but eyes-on approach is needed. A key factor in the success of partnering is continuous improvement and open-book accounting where cost and efficiency gains are auditable through access to the contractor's records.

1.3 Procurement methods and contracts in use

A wide variety of contracts are used for construction projects, including a number of standard forms. Every 2 years since 1985, the Royal Institution of Chartered Surveyors (RICS) has commissioned a survey to determine trends in the use of standard forms of contract and methods of procurement in construction. The survey is a representative sample of new build and refurbishment work carried out in the UK excluding civil engineering and routine maintenance work.

'Contracts in Use' (RICS 2003) found that a standard form of contract was used on 95% of building projects (86% by value). JCT standard forms were by far the most popular, accounting for 78% of contracts by value. There appears to have been a poor uptake of the NEC Engineering and Construction Contract for building work, despite it being heavily promoted by Latham. However, indications from the NEC User Group and other sources are that there is a more widespread use of the ECC in the civil engineering sector and in a number of countries abroad.

The RICS survey indicates that there has been a continuing sharp fall in contracts based on traditional bills of quantities; the number of these in use has halved since 1985. The use of drawings and specification has doubled over the same period.

Procurement trends seem to indicate a levelling off in the use of design and build since 1998, although this still represents 42% of the market by value. Construction management is on the increase (nearly 10%) but management contracting is in sharp decline (2%). The survey

PROCUREMENT METHODS

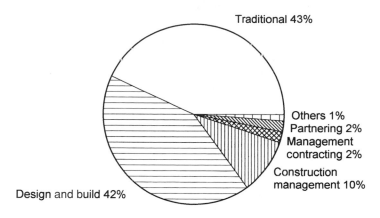

Traditional 43%

Others 1%
Partnering 2%
Management
contracting 2%

Construction
management 10%

Design and build 42%

By value of turnover
Excluding civil engineering, maintenance and repair work
Source: RICS

FORMS OF CONTRACT

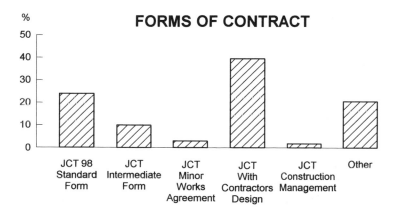

By value of turnover
Excluding civil engineering, maintenance and repair work
Source: RICS

USE OF PROCUREMENT AND CONTRACTS

Figure 1.1

includes figures on partnering but there are no trends evident except
that the use of binding and non-binding (charter) partnering agreements
are equally popular. It appears that binding partnering agreements
are more widely used on larger value projects. Figure 1.1 illustrates
the relative popularity of different procurement methods and standard
forms of contact.

In a private survey conducted by the authors, a large contractor with
a turnover of £300 million recently received the following types of
enquiry over a 6-month period:

- JCT Traditional   45%
- JCT With Contractor's Design   40%
- Partnering   10%
- Other (including management type)   5%

1.4 The Housing Grants, Construction and Regeneration Act 1996

The Act is in five parts and the Latham provisions are in Part II – Construction Contracts under sections 104–117. This legislation applies to all construction contracts whether main or subcontracts and also contracts for design or other professional services.

The Act confers certain statutory obligations and entitlements on parties to such construction contracts and where suitable clauses are not present in a contract they are provided by regulations called the Scheme for Construction Contracts. The Scheme represents the back-stop where perhaps standard conditions of contract are not employed or where contracts are entered into on the basis of an exchange of letters.

The issues covered by the Act include:

- The right to refer disputes on construction contracts to immediate adjudication rather than having to wait for arbitration or litigation
- The right to stage payments for work carried out under a construction contract
- The right to proper arrangements for payment including amount, dates and how it was calculated
- The obligation to give notice of intention to withhold payment including how much and why
- The right to suspend performance of the contract pending payment
- The prohibition of 'pay when paid' provisions (e.g. main contractor delaying payment to a subcontractor pending payment from the employer)

Most standard forms of contract have been amended to bring them into line with the Act. In particular, the JCT forms were changed to include the right of interest on late payments and the right to suspend work, subject to notice, if payment is not received on time.

1.5 Standard Forms of Contract

The Standard Form of Building Contract JCT 98

The JCT Standard Form with Quantities is a lump sum contract based on bills of quantities. A remeasurement contract is possible using either the approximate quantities or without quantities versions. The contract assumes separation of design and construction and the architect is designer, client's agent and contract administrator.

Partial contractor design is possible using the Contractor's Designed Portion Supplement in which case the contractor provides a Contract Sum Analysis of contractor's designed portion.

Under the contract, an Information Release Schedule may be provided stating what information the architect will release and when (this is optional for the client).

The contractor is to supply copies of his master programme (if he has one) but the programme does not impose any additional obligations on the contractor other than those determined by the contract documents. As the programme is not a contract document, failure to observe the programme is not a breach of contract.

A Sectional Completion Supplement is available for completion of the work in phased sections (as opposed to partial handovers at the contractor's discretion).

The architect determines extensions of time under this contract and certifies monthly payments normally based on interim valuations by the client's quantity surveyor who is named in the contract. Interim payments include materials on site and there are complex rules for the valuation of variations which are also paid as the work proceeds. Interest is payable on late payments.

Subcontractors can be domestic, nominated or named.

The contractor is to provide at full time competent person in charge and, when given possession of the site, he is to commence the works and proceed regularly and diligently in order to complete on or before the completion date.

Obviously, even the best laid plans rarely work out exactly but the only obligation to submit a revised programme under JCT 98 is where an extension of time is granted under the contract. The architect/contract administrator has no powers to require the contractor to accelerate the works but he may issue a notice of determination if the contractor persistently fails to progress satisfactorily in accordance with the intended programme.

The JCT Intermediate Form IFC 98

This is a popular form of contract for work which is straightforward, where there are no complex services and where nomination of subcontractors is not required. Subcontractors can be named by the contract administrator. In most respects the contractual provisions are similar to the JCT 98 Standard Form.

There is no contractual requirement for a programme under JCT 98. Therefore, where the contractor is required to produce a programme for the project, perhaps in a specified format, this should be stated in the bill of quantities, specification or preambles.

The Standard Form of Building Contract With Contractor's Design WCD 98

This contract is used for lump sum design and build contracts where the basis of the contract is Employer's Requirements and Contractor's Proposals. The contractor is to supply a Contract Sum Analysis of his price. The Employer's Agent is named in the contract but there is no named quantity surveyor.

Interim payments are based on the contractor's application either at monthly intervals or on completion of prescribed stages of work. Materials on site are not included in interim valuations where certificates are based on stage payments.

Unlike JCT 98, the design and build form contains no provision for a master programme. However, provision is made for determination of the contractor's employment under the contract for failure to proceed regularly and diligently with the works.

It would therefore be prudent for the client's advisors to have a copy of the contractor's up to date programme so as to monitor performance. This could be provided for in the contract preliminaries or in the instructions to tenderers. Alternatively, a requirement for a programme could be included in the Employer's Requirements which are incorporated in Article 4 of Appendix 3 to the Conditions of Contract.

The contractor's programme may well be an important document as payment in design and build may be linked either to construction stages or to activities on the programme.

The JCT Major Project Form of Contract 2003 MPF 03

This relatively new standard form of contract represents a significant departure from conventional JCT thinking and contains a number of novel features more reminiscent of the New Engineering Contract style of contracts. It is, for instance, written in a much shorter and simpler style and avoids the extensive procedural provisions associated with conventional JCT contracts.

As its name suggests, the Major Project Form is intended for use with experienced clients, contractors and subcontractors on large commercial projects where the client would otherwise wish to draft a bespoke contract or extensively amend an existing standard form. 'Major Project' is not defined but it seems the intention is to exclude 'run of the mill' projects for inexperienced clients.

The contractor undertakes more risk and responsibility than under conventional JCT contracts and it is essential that the contractor is well versed in the principles and application of risk management. The contract assumes that the client will have in-house procedures and protocols dealing with day-to-day matters such as instructions, record keeping, delays, updating the programme and health and safety monitoring, etc. that will be familiar to the contractor. Such detailed procedures, familiar to users of the main JCT forms of contract, have been much simplified in the Major Project Form and a certain degree of reliance is placed on the contractor having worked for the client on previous projects.

Some of the key features of the Major Project Form include:

- The employer describes the project in Requirements
- Consultants, subcontractors and specialists may be named in the employer's Requirements but there is no provision for nomination
- Certain provisions of the contract may be exercised by funders, tenants or purchasers via the Third Party Rights Schedule
- The contractor's Proposals show how he intends to meet the Requirements
- The Pricing Document sets out the means by which the contractor will be paid
- The contractor undertakes any further design beyond that given in the Requirements
- The contract includes a design submission procedure
- Payments for work in progress may be by interim valuation, stage payments or may be related to progress or, alternatively, may be another method (e.g. by S curve or formula)
- There is no provision for deduction of retention from interim payments
- The contractor undertakes the roles of planning supervisor and principal contractor under the Construction (Design and Management) Regulations 1994
- The contractor is given access to the site rather than possession of it as is conventionally the case on the assumption that the client may require others to carry out work on the site
- Unexpected ground conditions are dealt with in a similar manner to Clause 12 in the ICE Conditions of Contract
- The contractor is encouraged to suggest changes in order to make cost savings or value improvements
- There is no requirement for the contractor to provide a master programme
- Provisions are included to allow acceleration of the project by agreement between the parties
- Provision is made for bonus payments to be made for early completion
- The project may be carried out in stages without the need for a sectional completion supplement

The Major Project Form has been drafted by a firm of consultants, rather than the Joint Contracts Tribunal itself, under the direction of a steering group mainly representing the Construction Confederation and the British Property Federation. It is a much shorter form of contract than the JCT With Contractor's Design Form which it is likely to replace in due course.

The JCT Construction Management Contract CM 02

Under this standard form, the contract is between the client and the construction manager who is effectively on the client's team. The client appoints a consultant team and consultant team leader, and the construction manager has a limited role as client's agent with limited powers to spend money (maximum £15 000 per month – optional). The construction manager has a contractual obligation to consult with the consultant team on relevant matters.

The construction manager's obligations include:

- Recommend economies in buildability, methodology, cost and time during design development
- Prepare the project programme showing critical path, lead times and key milestones agreed with the client
- Prepare a detailed week-by-week programme listing all trade contracts and showing how the project is to be achieved
- Prepare a tender events schedule for each trade element
- Prepare trade contractors' detailed programmes and any acceleration proposals
- Prepare drawing and information release schedules showing procurement periods
- Advise on the division of the project into trade contracts
- Prepare and agree a suitable tender list and tendering arrangements
- Interview all tenderers before and during tender period
- Analyse tenders and recommend contractors
- Advise on the pre-ordering of materials and plant
- Submit monthly progress reports to the client
- Instruct, direct and manage trade contractors
- Check applications for payment from trade contractors and prepare valuations as necessary
- Make recommendations for payment
- Manage and coordinate the work of trade contractors
- Expand, update and adapt the project programme
- Monitor and report on actual expenditure against the project cost plan
- Organise and manage the site
- Maintain complete and accurate records

The Standard Form of Management Contract MC 98

Under these conditions, the contractor tenders on the basis of a management fee and works contracts are placed later for various 'packages' as and when the design is sufficiently developed.

The intention of this method of procurement is for the client to appoint professional advisors for the architectural and engineering design and quantity surveying (the professional team) and to engage a contractor in a 'professional' capacity to advise on design, technical and buildability issues.

In Article 1, the contract requires the management contractor to cooperate with the professional team during the design stages and in the planning, programming and cost estimating for the project. This involves the contractor in preparing the Project Programme which has to be agreed by the client's professional team.

Additionally, the management contractor is required to prepare all necessary programmes for the execution of the project, which includes the preparation of a detailed construction programme.

These obligations are emphasised in the Third Schedule of 'The Services to be provided by the Management Contractor'. The management contractor's duties include maintaining and regularly updating the detailed construction programme.

The ICE Conditions of Contract 7th Edition

This form of contract is used for works of a civil engineering nature or where building work is a minor part of the project. The contractor is required to submit his programme for the approval of the engineer, showing the order in which he intends to proceed with the works. This programme is commonly referred to as the Clause 14 Programme as it is required under clause 14(1) of the contract.

In addition to this obligation, the contractor is required to submit a written description of his proposed arrangements and methods for the project and, if requested by the engineer, detailed information on construction methods, temporary works and contractor's equipment. Additionally, the contractor is required to explain his intended methods of working and resourcing of the works to the reasonable satisfaction of the engineer, through clauses 13 and 14.

Submission of this information puts the engineer under an obligation to respond within a limited time by either accepting or rejecting the contractor's proposals or requesting him to submit further details. The contractor's programme is not usually a contract document in civil engineering but far greater emphasis is placed on the programme than in building works.

While a programme is not a contract document in its own right, the submission of a programme is a contractual requirement under clause 14. The engineer plays a significant role in its final production and he has the power to reject the proposed programme until such time as he is satisfied with the contractor's order of working.

Under ICE7, an up-to-date programme is a contractual requirement and clause 14(4) entitles the engineer to ask the contractor to submit a revised programme where actual progress is not as intended. Under clause 46(1), if the engineer considers that the contractor is not progressing in accordance with the programme, he is entitled to ask the contractor to expedite or speed up his work so as to complete the works on time. This does not, however, give the engineer the power to ask the contractor to accelerate the works so as to finish earlier than the contract completion date.

The provision for acceleration in clause 46(3) does not have the same meaning as 'expedite'. Acceleration provides a means for the client (employer) to ask the contractor to finish the job before the contract completion date (or the extended date), when perhaps an opening or commissioning deadline is important. In such circumstances the parties can agree special arrangements for earlier completion where mutually acceptable, but such an agreement will be outside the original contract.

The ICE Design and Construct Conditions of Contract

This contract is used for civil engineering works where the contractor is to design the works.

To all intents and purposes the contract is similar to the 7th Edition, with the obvious exceptions that apply to design and build arrangements. For example, the contract price is based on the Employer's Requirements and the Contractor's Submission. There is no engineer to the contract but there is an Employer's Representative instead.

The contractual obligation to provide a programme is similar to the 7th Edition.

The Engineering and Construction Contract (NEC)

The New Engineering Contract (NEC) was first published in 1991 and is now in its second edition following amendments recommended by the Latham Review. (A new edition is likely in 2004.) The NEC was considered by Latham to be capable of being a common contract for the whole industry both in the public and private sectors.

The main form of contract has been renamed The Engineering and Construction Contract (ECC) within the NEC family of documents which also includes standard forms of subcontract and professional services contracts, etc. These changes, however, have not altered the fundamental concept of the NEC which was designed to be flexible in its use and applicable to a variety of types of construction work including:

- Building
- Civil engineering
- Process and other engineering
- Refurbishment work
- Maintenance work

The contract can be used for both major and minor works and its flexibility extends to employing a variety of contractual arrangements with different types of project documentation.

A project can be set up on the basis of the main options shown in Table 1.2.

There is no architect or engineer under the NEC. Instead these traditional appointments are separated into the functional roles of project manager, supervisor and adjudicator, although of course they could be undertaken by one individual if preferred. The flexibility of this approach can be exemplified, for example, where a construction management approach is required. Here, the client would enter into direct contracts with each of the works contractors using the main ECC form and appoint the management contractor as 'project manager' under the same form of contract.

One of the basic concepts of the NEC is that any contract requires certain fundamental clauses, such as provision for time, quality, payment and dispute resolution. Other clauses, on the other hand, can be considered as 'optional extras'. These might deal with liquidated damages, bonds and guarantees, retentions, early completion bonuses, etc. Consequently, the NEC has nine 'core clauses' which would be in every contract, and fifteen 'secondary option clauses' (plus a partnering option)

Table 1.2 Main options of NEC.

Option	Brief description
A	A priced contract using an activity schedule (a list of activities linked to the programme)
B	A priced contract using bills of quantities (e.g. SMM7 or CESMM)
C	A target contract using an activity schedule
D	A target contract using a bill of quantities
E	A cost-reimbursable contract using schedules of actual cost (similar in principle to daywork)
F	A management contract (with works package contractors engaged directly by the management contractor)

from which to choose to bespoke the contract to the client's require-ments. The flexible approach offered by the NEC facilitates consultant or contractor design and also contracts based on conventional lump sum, remeasurement, management arrangements and target cost or other reimbursable contracts.

The programme is an important document under this form of con-tract and there are contractual obligations placed on the contractor under clause 30 prescribing information to be shown on the programme and to make sure it is kept up to date. The programme must show:

- Critical dates
- Operations by the contractor and others
- Equipment and resources for each operation
- Order and timing of the works
- Float and time risk allowances
- Health and safety requirements
- Other general information and dates

ACA Standard Form of Contract for Project Partnering PPC2000

This form of contract has been specifically designed by the Association of Consultant Architects for use on projects where a partnering arrange-ment is intended. It may be used in conjunction with a partnering charter which is an agreed statement of values, goals and priorities signed by the partnering team members. The key features of the contract are:

- A partnering team who are the signatories to the partnering contract
- A contractual undertaking by the partnering team to work together in a spirit of trust, fairness and mutual cooperation to the benefit of the project
- A client representative acting as fully authorised client's agent (except membership of the core group)
- A partnering advisor
- A core group drawn from the partnering team which meets regu-larly to arrive at consensus decisions which are binding on the partnering team
- A list of six partnering objectives and ten partnering targets
- An early warning system whereby partnering team members notify others of potential problems
- A partnering timetable setting out the activities of the partnering team prior to commencement on site
- A project timetable for implementing the project post commence-ment agreement
- Implementation of an open-book approach to supply chain relationships

- Risk management exercises aimed at identifying, eliminating or reducing risks and their costs and apportioning risks to the partnering team members best able to manage them
- 16 provisions enabling the constructor to be granted extensions of time subject to his best endeavours to minimise delay and increased cost
- Monthly payment intervals, unless stated otherwise
- Payments subject to valuation by the client's representative
- Payment 15 days from valuation or 10 days from receipt of VAT invoice
- Payment under a pre-possession agreement for work done before commencement
- Payment in accordance with the price framework subject to an agreed maximum price

1.6 Tendering procedures

Tendering arrangements for construction projects are normally conducted according to the procedures laid down in various National Joint Consultative Committee (NJCC) Codes of Procedure for tendering for a variety of procurement arrangements including single-stage, two-stage and design and build. These codes are not mandatory but are widely observed in the industry and they provide guidance to good practice with respect to suitable tender periods, the number of contractors invited to tender and the general conduct of the tender process.

For public sector projects over certain value thresholds, UK tendering regulations, which implement EC directives, have to be followed. The rules vary according to the value of the project. Open, restrictive (selective) and negotiated tendering is allowed and there are proposals to include framework arrangements in a future consolidated EC public sector directive. Public sector projects over a certain value have to be advertised in the *Official Journal of the European Communities* (OJEC).

The effect of the public sector tendering regulations is to impose time constraints on tendering periods, and 'lead' times have to be allowed for receiving requests to tender (minimum 36 days) and for receipt of tenders (up to 40 days minimum). These have to be taken into account when deciding on the overall time-scale for public sector projects.

Twort and Rees (2004) give a good overview of tendering procedures and legislation.

1.7 Procurement options

When a client is considering the decision to build, a number of important decisions need to be made. The NEDO report 'Thinking about

Building' (NEDO 1985) suggested eight procurement factors to be considered by the client at the procurement stage. The three main considerations were:

- Time – the overall timing of the project from inception to completion
- Quality – the client's required standards of design and workmanship as expressed in the specification
- Cost – the market price applicable at the bid submission stage and the final account cost

Five other client considerations were also highlighted as being influential on the choice of procurement route. These were:

- Complexity – Complexity of building design, layout and services provision
- Controllable variation – How sure is the client of his requirements? The cost of variations to the contract may prove difficult to agree and assess
- Degree of competition – Does the client wish to create competition at the design and construction stages of the project?
- Client responsibility – Does the client wish to be directly involved in decision-making during the project or does he simply want to return from the South of France and cut the tape on occupation of the building?
- Risk in the project – Commercial risk, occupation risk, design and construction risk

Taking these factors into account, decisions will have to be made in relation to:

- Choosing an appropriate procurement strategy for managing the project
- Selecting the client's principal advisor or lead consultant. The 'butcher, baker or candlestick maker' approach may have to apply here, i.e. architect, engineer, project manager
- The appointment of other consultants
- Establishing the client's brief and assessing the project's feasibility
- Obtaining the required project finance
- Choosing how to procure and manage the design and construction
- Establishing a realistic time period for the overall project

A detailed overview of procurement assessment criteria is outlined by Turner (1997) and Seeley (1997).

The procurement of a construction project requires a balance between time, quality and cost constraints. A critical eye needs to be kept on the client's requirements and the forecast budget. This necessitates effective briefing, design and cost control in relation to the final choice of procurement route.

Procurement methodology concerns not only the contractual arrangement for the construction work but also the sourcing of professional services for design, project administration and health and safety management.

Turner (1997) considers who should carry the risk and how the right choice of procurement route can ensure that the risk is allocated to whichever party is best placed to manage it.

Procurement methods are constantly under review and whilst there may be a significant trend towards design and build arrangements (both client-led and contractor-led design), other procurement options such as construction management and partnering arrangements are coming to the fore. For large contractors operating in the public sector, the decision to tender for PFI contracts requires major decisions at senior management level because of the tendering costs and risks involved.

The following common procurement options available to clients are outlined in the rest of this chapter:

- Traditional procurement (JCT 98 standard contract with or without quantities)
- Design and build arrangements (JCT 98 standard contract with client or contractor-led design)
- Management contracting (JCT MC 98 Management Contract)
- Construction management (JCT CM 02 Construction Management Agreement)
- Project management
- Private Finance Initiative (PFI) or Public–Private Partnerships (PPP)
- Framework contracting
- Prime contracting
- Partnering arrangements (incorporated into the ethos of the contract)

1.8 Traditional procurement

Introduction

Despite the different procurement options available, the recent RICS survey of contracts in use suggests that, in the majority of cases, a traditional route will be adopted with the client's lead consultant being an architect/designer. The RIBA Outline Plan of Work (RIBA 2000) is a model framework for running a construction project on this basis, both for managing the job itself and for running the necessary office procedures. Procedures are laid down both for a conventional fully designed project and for situations where there may be 'employer's requirements' and 'contractor's proposals' (i.e. design and build), in which case the architect may subsequently be novated to the contractor-designer.

Alternatively, a project manager may be engaged as the client's representative. The project manager may then decide to follow the project development procedures recommended in the CIOB Code of Practice for Project Management (CIOB 2002).

In any event, the lead consultant will interpret and develop the client brief, prepare cost and feasibility studies, arrange for scheme drawings to be finalised and organise the tenders.

Forms of contract

The JCT Standard Form of Contract (with or without quantities) may be used but other forms such as the JCT Intermediate Form or the JCT Minor Works Agreement may be applicable for less complex or small projects. The New Engineering Contract may also be considered, with the client representative acting as 'project manager', i.e. both lead designer and contract administrator. For civil engineering works, the ICE Conditions would normally be used with the engineer in the role of client representative.

An alternative is to appoint a project manager to represent the client as lead consultant, with the design team, including the architect, under his direction. This would require careful consideration, especially in terms of the form of contract used, as only the ECC gives the opportunity to separate the lead design and contract supervisor roles in the contract. Other contracts would need to be amended because, for instance, the JCT Standard Form refers to the 'architect' and the Intermediate Form to the 'architect/contract administrator'.

The JCT contract has provisions which cater for variations, extensions of time and loss and expense (clauses 13, 25 and 26), but it is arguable whether contractors recover the true cost of delay and disruption to the work should this happen. These difficulties are usually handled by most contractors with good humour and a ready acceptance of a reasonable reimbursement, but frustration sets in when the client refuses to agree or pay for the extra work. Litigation often results and brings to an end a reasonable relationship which the parties worked hard to establish during the project.

Project relationships

Figure 1.2 illustrates the relationships between the client, client's design team and the main contractor in traditional procurement. Figure 1.3 shows the work stages from the RIBA Plan of Work which the architect

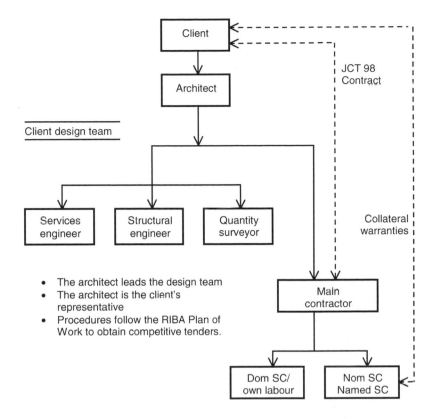

BUILDING CONTRACT
(JCT 98 With Quantities)

TRADITIONAL PROCUREMENT RELATIONSHIPS

Figure 1.2

follows from contract feasibility stage through to completion of the project. The work stages are broadly classified under three main development stages in a project:

- Feasibility – stages A–B
- Pre-construction – stages C–H
- Construction – stages J–L

Figure 1.4 indicates the relationships between the parties on a civil engineering project using the ICE Conditions of Contract 7th Edition. The client's lead consultant is normally a consulting engineer who will be responsible for coordinating the design and construction process. Works on site are usually under the direction of a resident engineer assisted by an inspector of works to ensure quality and workmanship.

RIBAOUTLINE PLAN OF WORK
TRADITIONAL CONTRACT

Procedures undertaken by an architect
during the designing and administration of a building project

STAGE	COMMENTS	
A – APPRAISAL	Confirmation of client's requirements, identify procedures. Engagement of consultant and select procurement route.	FEASIBILITY
B – STRATEGIC BRIEFING	Preparation of client's brief. Confirm key requirements. Identify procedures. Organise structure of design team.	
C – OUTLINE PROPOSALS	Develop full brief. Prepare outline proposals. Estimate of cost. Review procurement route.	PRE-CONSTRUCTION PERIOD
D – DETAILED PROPOSALS	Complete brief. Prepare detailed proposals. Apply for development control approval.	
E – FINAL PROPOSALS	Preparation of final proposals, scheme plans, elevations. Coordination of all components and elements.	
F – PRODUCTION INFORMATION	Preparation of production information to enable tender documents to be prepared. Application for statutory approvals.	
G – TENDER DOCUMENTATION	Preparation of tender documentation. Tenders to be invited for project.	
H – TENDER ACTION	Evaluation of tenders received. Tender comparisons. Appraising tenders and recommendation to client.	
J – MOBILISATION	Award of building contract, appoint contractor. Issue production information. Occupation of site by contractor.	CONSTRUCTION PERIOD
K – CONSTRUCTION STAGE	Administration of contract to practical completion stage. Provision of contract information during project.	
L – AFTER COMPLETION	Administration of contract after completion. Final inspections, settle final account.	

Figure 1.3

Advantages of traditional procurement

Despite the criticisms, traditional procurement has the advantage of price competition from competing contractors at the tender stage and should work successfully provided that:

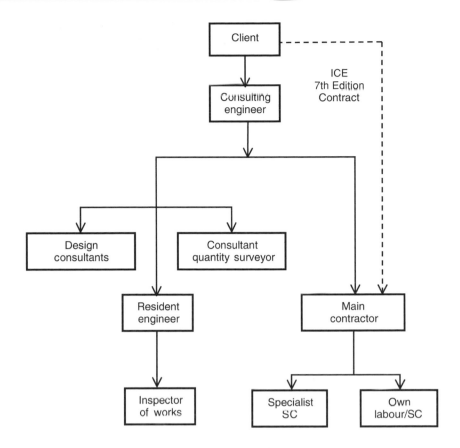

CIVIL ENGINEERING PROJECT
TRADITIONAL PROCUREMENT RELATIONSHIPS

Figure 1.4

- The design is complete before the tender stage (ensuring price certainty for the client)
- The designer understands how the building 'fits together', ensuring buildability
- The design does not substantially change during construction, so contract variations are kept to a minimum

Summary

- Client knows the lump sum cost before being committed
- Firm contractual date for completion
- Client retains control over the design team and quality can be assured

- Capable of obtaining the best contract price for the full scope of the works
- Variations and contract changes are relatively easy to handle (but often not so easy to agree!)

Disadvantages of traditional procurement

At its best traditional procurement is ideal for many clients, but all too frequently:

- The end product is not finished on time
- The budget is overspent as the final account far exceeds the client's expectation
- Disputes arise
- Adjudication, arbitration or litigation results with the only winner being the legal profession

Summary

- Commitment to a lump sum price is often undermined by a lack of information or client changes
- Tenders can only be finalised when the design is complete, which may lead to an extended programme
- Responsibility for specialists lies with the client
- All design risk is ultimately carried by the client
- High pre-contract design fees likely

1.9 Design and build

Introduction

The work stages outlined in the RIBA Plan of Work for the traditional contract have been extended to include design and build arrangements. The Outline Plan of Work (RIBA 2000) includes sections for the procurement of the Employer's Requirements (Part 3) and the Contractor's Proposals (Part 4).

The work stages have the same work stage titles but some of the descriptions reflect the particular nature of the employer and contractor requirements. The work stages relevant to the input from the client's consultant team (architect, civil/structural engineer and services engineer) will depend upon the extent to which the client wishes to influence the design and construction detail.

Two extremes may be considered in relation to design and build procurement, Extreme A and Extreme Z.

Extreme A – Client-led design and build

In this situation, the client wishes to be fully involved in influencing the design. Prior to tender action (Work Stage H), the design will be virtually complete and may include full bills of quantities or notional bills. At this stage a number of contractors will be invited to tender for the project. Here, the design risk is taken fully by the client.

The client will be responsible for all design fees from commencement to completion of the project. The architect may be appointed lead consultant or the client may decide to appoint a project manager in the lead consultant's role. Figure 1.5 illustrates the possible relationships between the client, design team and contractor.

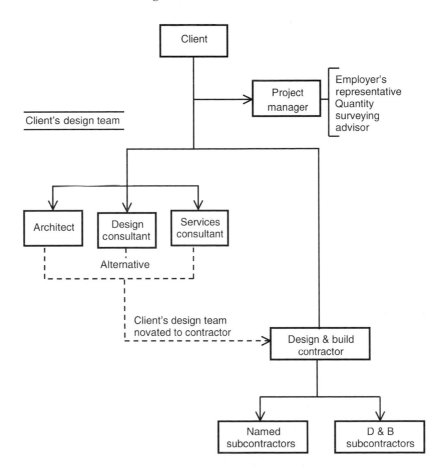

**DESIGN AND BUILD
CLIENT- LED DESIGN RELATIONSHIP**

Figure 1.5

An alternative consideration often used by the client is to novate the design team to the contractor once the contract has been awarded. In this way the client will maintain an interest in the design but the contractor will pay the continuing design fees through to contract completion. Where novation is chosen, it is advisable that the client's project manager has a consultant quantity surveyor as part of the client's representation during the project.

Extreme Z – Contractor-led design and build

In this situation, the client may wish to provide the contractor with minimal information in the form of an outline brief (RIBA Stage B – Strategic Briefing). This will leave the design and build contractor fully responsibile for the conceptual and detailed design in order to meet the Employer's Requirements.

The contractor will then submit a bid based on the Contractor's Proposals. Full design responsibility will be taken by the contractor to produce a building in respect of the client's time, quality and cost requirements.

Figure 1.6 illustrates the relationship between the parties. The contractor may provide an in-house design facility or independent design teams may be used. With a team arrangement, it is normal for the design and build contractor to employ a design team coordinator as a key member of the team, to ensure the flow of information between the design team and the project team in order that key design and construction dates are adhered to.

Other design and build options (somewhere between Extremes A and Z)

It is most probable when using design and build procurement that procedures will fall somewhere between the two extremes described above. Definitions of alternative arrangements are outlined by Janssens (1991) (see Table 1.3) and these indicate the wide range of options available.

Table 1.3 Design and build arrangements.

Extreme	Variety
A Employer-led design	Develop and construct
	Design and build (single-stage tender)
Z Contractor-led design	Design and build (two-stage tender)
	Negotiated design and build
	Design and manage
	Turnkey

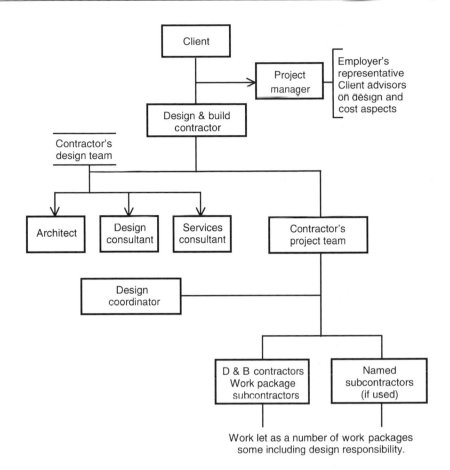

DESIGN AND BUILD
CONTRACTOR – LED DESIGN

Figure 1.6

The main difference between the options in Table 1.3 usually relates to the proportion of design work undertaken by the employer's consultants. This may simply be a conceptual design or something more developed. It is this design that is included in the tender enquiry to the contractor, from which the contractor's proposals are developed.

Advantages of design and build

- Price certainty is secured early in the project
- Contractual completion dates are fixed early in the design process
- Less risk of price changes during the design development

- Main contractor provides single-point responsibility for design and construction
- Major risks lie with the contractor

Disadvantages of design and build

- Contractor assumes greater financial risk and this is often reflected in the price
- Difficulties arise for the employer in matching like-for-like prices at tender stage. For example, the client's requirements indicated that he wanted a lemon tree. Some contractors priced for an orange tree and others simply priced for a tree
- Tender period and negotiation tends to be much longer
- More problematic to control design and quality
- Changes can prove to be expensive and disruptive to the contractor
- Project becomes price driven at the expense of quality

1.10 Management contracting

As an alternative to the traditional approach the client may decide to appoint a management contractor to engage and manage a number of work package subcontractors.

Management contracting enables the client to create a competitive situation between management contractors at the appointment stage of the project. For instance, four management contractors may be invited to submit quotations for managing the project and would then present their proposals to the client and his design team. It is important that the design team feel that they can 'gel' with the management contractor and work as a project team. The presentations would indicate the management contractor's approach to:

- The programme
- The procurement of work packages
- The control of information flow
- Buildabilty considerations

Bids are based on the contractor's assessment of management and preliminaries costs associated with managing the work packages. Further competition is created during the letting of the work packages as each work package subcontractor bids competitively to obtain the work. Certain work packages may contain a design element. Figure 1.7 illustrates the relationships between the parties involved in the project.

The work package approach allows maximum overlap between

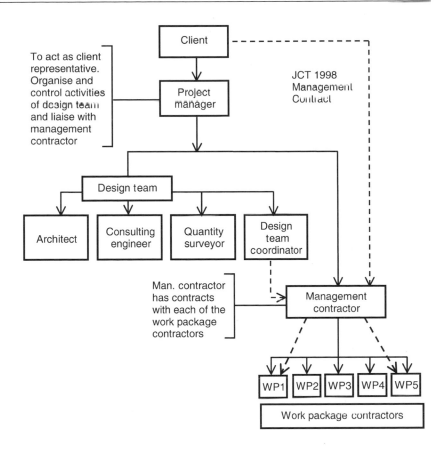

MANAGEMENT CONTRACTING ARRANGEMENT

Figure 1.7

design and construction activities. As each section of work is designed, the work package subcontractor may commence the works on site. This tends to result in a reduced overall contract period; if only it worked consistently in practice, it would be wonderful.

Strict control of the work package budget is essential for the success of the project and many prestigious projects have shown the tendency to overspend. Problems in practice have also resulted from the management contractor failing to pay the work package subcontractors in due time. When valuation payments are sometimes in the order of £10 million per month, a 1-week delay in paying the subcontractors will certainly help to improve the management contractor's cash flow. Management contractors have even been known to use the 'payment threat' as a means of getting the work package subcontractor to perform.

1.11 Construction management

Figure 1.8 illustrates the relationship between the parties in construction management procurement. In this situation, the construction management consultant or construction manager joins the design team early in the project.

The construction manager's services are based on a negotiated fee with the client simply to supervise and plan the work to be undertaken by the work package contractors. Normally, this method of procurement is reserved for major projects but on a small factory/office block project, the construction manager could simply be a self-employed person offering a management service to supervise and coordinate a number of subcontractors.

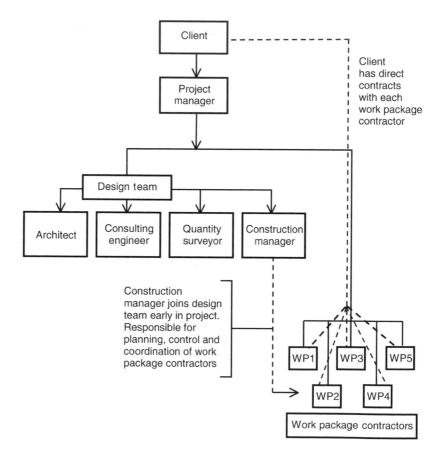

CONSTRUCTION MANAGEMENT ARRANGEMENT

Figure 1.8

Each work package contractor has a separate contract with the client and is paid directly by the client each month.

Advantages of construction management and management contracting

- Maximum overlap between design and construction. As each package design is completed work may commence on the project
- Quality can be assured – a high level of supervision and quality control can be achieved at site level
- The construction expertise of the management contractor aids buildability and value engineering
- Variations can be kept under control as the price of changes can be agreed before commencing a variation
- The contractor takes a less adversarial approach, being part of the client's team (he has learnt all the 'dodges' by practical experience)
- Suits complex projects where the design can be developed in stages

Disadvantages of construction management and management contracting

- The client has no commitment from the contractor on price certainty. Strict control of the work package budget is essential
- Total cost of the project is not usually known until the project is well into the construction programme
- Blue chip work package subcontractors are often chosen with no incentive to reduce costs
- Early letting of packages may lead to extensive design changes
- Damages for delay are difficult to pin on one subcontractor and they are expensive to negotiate
- The client takes all the risk, particularly in the construction management arrangement

1.12 The role of the project manager

Definitions

Walker (2002) suggests that:

- Management is the dynamic input that makes the organisation work
- Organisation is the pattern of the interrelationships, authority and responsibility that is established between the contributors to it

More pragmatically, the Chartered Institute of Building Code of Practice for Project Management (CIOB 2002) provides clients and all members of the construction team with a comprehensive management strategy for any project from inception to completion.

The Code defines project management as 'the overall planning, control and coordination of a project from inception to completion aimed at meeting a client's requirements in order to produce a functionally and financially viable project that will be completed on time within authorised cost and to the required quality standards'.

Time, cost and quality control are the keystones to the success of a project which many project managers find most difficult to achieve.

Under the direction and supervision of a project manager, management of the project becomes a team effort, as shown in Figure 1.9. The duties of the project manager and his formal agreement and conditions of engagement with the client are indicated in the CIOB Code.

PROJECT MANAGEMENT – TEAM STRUCTURE

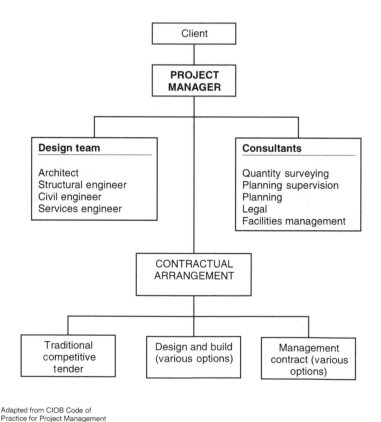

Adapted from CIOB Code of
Practice for Project Management

Figure 1.9

Confusion often arises in the use of the title 'project manager'. Walker suggests that the client's project manager may be classed as an executive or non-executive project manager. The non-executive project manager is effectively a team coordinator who may have another role within the team (such as design). The executive project manager, on the other hand, is independent from the other members of the team and is purely concerned with managing their activities.

The contractor may also use the title 'project manager' to describe his senior construction representative on site and this often implies that he is a super-duper site agent. This anomaly often confuses site and management personnel in the use of individual titles.

Confusion also arises with the use of the term 'client's agent', which may imply either an executive or non-executive role. There are further complications in terms of the client's agent under the Construction (Design and Management) Regulations 1994 who undertakes the client's statutory duties either fully (if notified to the Health and Safety Executive) or partially (if not notified).

Project management stages

Figure 1.10 illustrates the main stages involved during a project from inception to completion. This process somewhat mirrors the RIBA Plan of Work where the architect would take the role of lead consultant (non-executive).

Figure 1.11 is based on the project management handbook of a local authority which has established procedures for taking a project management approach to managing new build projects. This model has proved extremely successful with the authority due to the simplicity of its application to projects.

1.13 Private Finance Initiative (PFI)

PFI is an arrangement where public sector assets and services are acquired through private sector funding with the idea of reducing government/public sector borrowing. PFI is one of several types of public–private partnerships (PPPs) and typical projects undertaken include roads, bridges, schools, hospitals and prisons. Projects are typically over £20 million in value.

Once the public sector sponsor has established a business case strategy, the project will be advertised in the *Official Journal of the European Community* (OJEC) and pre-qualified bidders will be short-listed. Bidders for PFI contracts will invariably be a joint venture or consortium of firms (contractors, or contractors and design consultants)

DEVELOPMENT PROCESS AND THE PROJECT MANAGER

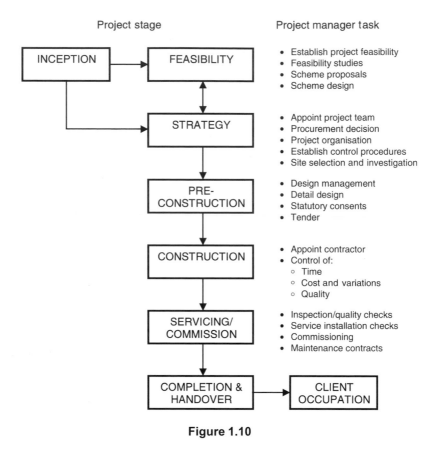

Figure 1.10

who will set up a Special Purpose Company (SPC) especially for the project.

The SPC will bid for the project in competition and, following a costly and lengthy process, a preferred bidder will be chosen. This point is reached through a series of gateways which punctuate the process at key stages. The contract will then be awarded and notified in the OJEC.

The project will normally be financed by borrowings from a bank or investment company at a high rate of gearing. Typically, the SPC members will provide equity of 10% and the remainder will be borrowed. This is risky for the project sponsor (client) and therefore a step-in agreement may be arranged which will enable the project to continue if the SPC fails. Subcontractors on site are particularly at risk as they are effectively unsecured creditors of the SPC whose only assets are the project and any equity funds remaining. The complex contractual and funding relationships are simplified in Figure 1.12.

LOCAL AUTHORITY MODEL
IMPLEMENTATION OF PROJECT MANAGEMENT

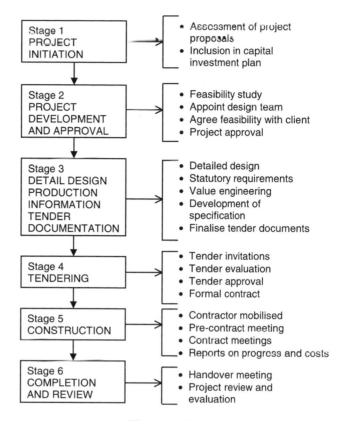

Figure 1.11

Most PFI projects are design–build–operate schemes with capital invested coming back over a 25–30 year concession period where tolls (e.g. roads) or service charges (e.g. schools) are made. Some schemes are partly financed by the public sector.

Examples of PFI projects include the Dartford and Skye Bridge crossings, the Channel Tunnel and the Second Severn crossing, which operate under a DBOOT (design–build–own–operate–transfer) arrangement.

Advantages of PFI

- Potential for high returns
- Continuity of work
- Involvement in design
- Buildability input
- More control over the programme

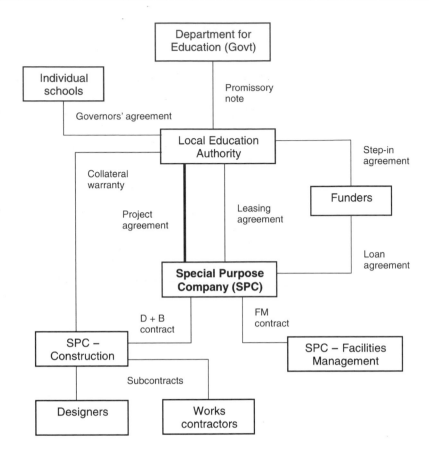

**PFI RELATIONSHIPS FOR A SCHOOLS PROJECT
WITH PARTNERING**

Figure 1.12

Disadvantages of PFI

- Very high bidding costs
- Competitive market
- Long bidding process
- High level of resources required
- Complex and demanding
- Tough contract terms
- High level of liquidated and ascertained damages
- Fixed price risk to contractor

1.14 Framework contracting

In recent years 'traditional' approaches to tendering and contractor selection have been found to be defective, as reported by Latham and Egan for example. In this respect, final delivery of projects based on the criteria of selection by price has been found to result in poor quality and unreliable outcomes and this is why other approaches have been devised.

One of these is framework contracting. This idea is similar to the traditional standing tender list in that approved suppliers are chosen or preferred to go on the list for future contracts. The framework suppliers include contractors, specialist contractors, suppliers of goods and suppliers of professional services (such as architects and surveyors) who are invited to tender for packages of work.

The big difference, though, is that with a framework contract there will be no 'main contractor' as such. Instead, there will be a number of first tier framework suppliers tendering for the main packages of work, and possibly second tier suppliers tendering for the smaller or specialist packages.

Another difference is that the choice of successful bidder is not made solely on the basis of price. In this way the framework of supervision, project management and construction is built up for a specific project that is then implemented on a sort of 'partnering' basis.

Big clients, such as British Airports Authority, use this system and their projects are run by the framework team and not one single 'managing' contractor as is traditional in the building industry.

1.15 Prime contracting

Prime contracting runs on similar lines to the standing approved lists but the prime contractors are chosen very carefully, usually via several stages of selection. The system is used by some Government departments such as Defence Estates and the National Health Service.

The idea is to have prime contractors who can provide all the project deliverables for capital projects, including design, planning and cost control. There are also prime contractors or one-stop shops who are able to deliver all property maintenance and capital works in particular regions. Once a prime contractor is chosen, he becomes part of the client's project team for perhaps up to 7 years. The client then takes a hands off–eyes on approach to their projects but with 'partnering' very much as the focus.

Selection is based on 'hard' and 'soft' criteria. Hard criteria are the usual ones of price, competence and standing, and the soft criteria focus largely on ideas, attitudes, willingness to share risk, flexibility, etc.

Features of prime contracting include:

- The point when a price is agreed with the prime contractor can vary
- The price could be fixed price, guaranteed maximum price (GMP), target price, etc.
- The form of contract could vary
- The period of facilities management is flexible
- Long-term commercial agreements could be built in for facilities management, other contracts, etc.

1.16 Partnering

Definition of partnering

Partnering may be defined as 'a structured methodology for organisations to set up mutually advantageous commercial arrangements, either for single projects or in long-term strategic relationships, which help their people work together more effectively' (CIB 1997).

It should be stressed that partnering is entirely a voluntary arrangement which operates within the chosen procurement method. The essential components of partnering are illustrated in Figure 1.13 and these components are fundamentally geared to establishing a commitment from participants to work together for mutual success where:

- All participants seek 'win–win' solutions
- Value is placed on a long-term relationship

PARTNERING

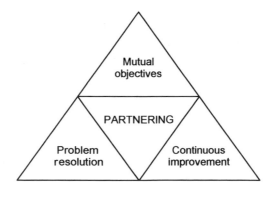

Source: The Reading Construction Forum – 'Trusting the Team'

Figure 1.13

- Trust and openness are norms
- An environment for long-term profitability exists
- All are encouraged to openly address any problems
- All understand that no-one benefits from exploitation of the others
- Innovation is encouraged
- Each partner is aware of the other's needs, concerns and objectives and is interested in helping their partner achieve this
- Overall performance is improved

Partnering arrangements

Partnering arrangements are quite normal in construction and a number of close client/contractor relationships have developed over the years. One of the best known 'alliances' is between Marks and Spencer and Bovis, although this has never developed into true or exclusive 'partnering' and other contractors also tender for and work for this particular client.

There are two types of partnering arrangement:

- Strategic partnering
- Project partnering

Strategic partnering is where a relationship is developed over an indefinite period and where there is a long-term commitment to the partnering approach. However, partnering may be for a particular project or length of time, such as for an individual contract or for a three-year maintenance contract. This would be a 'project partnering' approach.

The partnering arrangement may be based on a formal contractual arrangement between the partners, such as the PPC2000 Form of Contract for Project Partnering or the Engineering and Construction Contract using secondary Option X12. Alternatively, there can be a non-legally binding relationship based on a charter with a mission statement and a set of common objectives such as the JCT Non-binding Partnering Charter for a Single Project. In either case there would need to be a basis for agreeing prices and conditions of contract that are fair to both parties and this is left to the discretion of the contracting parties as normal.

Partnering charters are common and a charter may even be included in a formal partnering contract (see PPC2000, clause 5.6). The contents of partnering charters vary but might contain some or all of the following objectives and agreements:

- Budget certainty for the client
- Agreed maximum price
- Positive cash flow for the contractor

- Reasonable profitability for the contractor
- Completion on time
- A good/exemplary safety record on site
- Good relationships
- Enhanced reputation
- Trust and fairness
- Encouragement of ideas and innovation
- Sharing of cost savings
- Early warning of problems (two-way)
- Joint problem solving
- Use of conciliation/alternative dispute resolution where all else fails

A partnering culture includes:

- Open communication
- Top-down commitment
- Honest dealings (the difficult bit!)
- Elimination of confrontation
- A problem-solving approach
- Proactive teamwork

The partnering team will consist of all supply-chain participants in the project and will include the customer and end-user.

Partnering in practice

Project partnering is most common in the UK but both strategic and project specific relationships need to be based on trust, dedication to common goals and an understanding of each other's individual expectations and values. Expected benefits include improved efficiency and cost-effectiveness, increased opportunity for innovation, and the continuous improvement of quality products and services.

Partnering arrangements may be preferred where the client has repeat business for the industry or perhaps wishes to adopt a Latham/Egan or team-building approach to a specific project. This can be achieved by entering into a relationship over a specified period or for a longer period, where both parties agree to work together to achieve particular goals, for mutual benefit. However, any arrangement that does not have the triangle of features in Figure 1.13 is an 'alliance' and not true partnering.

Figure 1.14 illustrates the partnering process using a traditional procurement route but the principles of open discussion and consensus using workshops or 'shirt-sleeved' meetings are common whatever the preferred procurement arrangements.

EXAMPLE OF CLIENT/CONTRACTOR PARTNERING USING TRADITIONAL PROCUREMENT METHODS

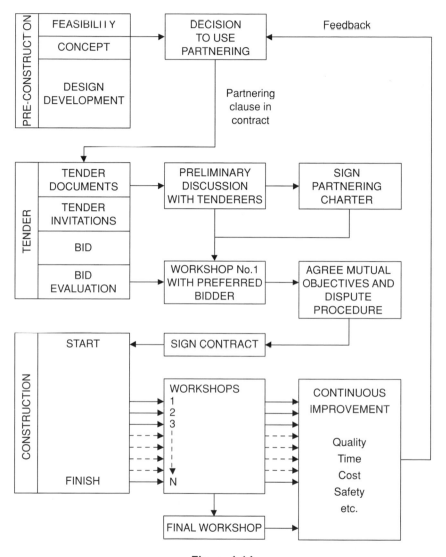

Figure 1.14

The benefits of building up long-term relationships were recognised in the Latham Report in terms of improving quality and timeliness of completion while reducing costs. However, Latham also recognised that the benefits need to be mutual for the system to work and, whilst contractors may be assured of continuity of work, clients must be assured that prices remain competitive within a relationship of trust.

Feedback from completed projects suggests that typical benefits from using a partnering approach include:

- Cost savings through value engineering
- Fewer interest charges due to no retention
- Cost savings through buildability
- Cost savings by eliminating formal tendering
- Reduced site correspondence
- Good relationships leading to repeat business
- Effective cost management of variations
- Zero accidents

References

CIB (1997) Construction Industry Board. *Partnering the Team*. Thomas Telford.

CIOB (2002) Chartered Institute of Building. *Code of Practice for Project Management*, 3rd edn. Blackwell Publishing.

Construction Task Force (1998) *Rethinking Construction*. Department of Trade and Industry.

Janssens, D. E. L. (1991) *Design-Build Explained*. Macmillan Education.

Latham, Sir M. (1994) *Constructing the Team*. HMSO.

National Audit Office (2001) *Modernising Construction*. Report by the Comptroller and Auditor General HC 87. The Stationery Office.

NEDO (1985) *Thinking about Building*. National Economic Development Office. HMSO.

RIBA (2000) The Architect's Plan of Work, Royal Institute of British Architects. RIBA Enterprises.

RICS (2003) *Contracts in Use – A survey of building contracts in use during 2001*. Royal Institution of Chartered Surveyors Construction Faculty.

Seeley, I. H. (1997) *Quantity Surveying Practice*, 2nd edn. Macmillan Press.

Turner, A. E. (1997) *Building Procurement*, 2nd edn. Macmillan Press.

Twort, A. C. & Rees, J. G. (2004) *Civil Engineering Project Management*. Elsevier Butterworth-Heinemann.

Walker, A. (2002) *Project Management in Construction*, 4th edn. Blackwell Publishing.

2 Management and organisation

2.1 Management principles

The principles of management established by Henri Fayol in the early twentieth century are as applicable today as they were then. The seven main principles are outlined by many management writers including Brech, Denyer, Drucker, Calvert, Cole and Clutterbuck.

Clutterbuck and Crainer (1990) provide a particularly useful overview of individual contributions to management development in a chronological order from 1841 to date.

The seven principles of management

Fayol developed seven basic principles which are generally applicable to a wide range of business organisations. The seven principles comprise:

- Forecasting and planning
- Organisation
- Commanding or directing
- Controlling
- Coordination
- Motivation
- Communicating, which encompasses them all

Each of these principles is now considered below as applied to the management of a construction organisation.

Forecasting and planning

Forecasting is looking into the future, and planning involves the making of decisions based on these forecasts. Types of planning at management level include considerations of strategy (strategic planning), business planning, and long and short-term business planning. Setting business objectives and policy making form an integral part of forecasting and planning.

Planning first involves consideration of the objectives of the business. A contractor's objective is to make a profit in relation to the amount of capital invested – no profit, no business. Second, policies must be established in relation to the planned rate of growth and expansion.

Forecasting involves the preparation of:

- Financial forecasts
  - o company annual turnover
  - o company cash funding requirements
  - o individual project cash funding
- Construction workload forecasts and estimating workload
- Resource forecasts
  - o Staffing and key labour
  - o Subcontractor resource

Organisation

Organisation is the grouping of work and the allocation of duties, responsibilities and authority.

- Organisation structure of company
- Organisation of individual functions (departments)
- Organisation structure of projects
- Defining roles and responsibilities within the organisation
- Providing job descriptions for individual staff

One of the steps in organising is to divide the business into sections which will be of most help in the efficient administration of the business. These sections may be related to functions, product or location. Many construction firms are organised functionally, i.e. surveying, buying, administration, planning and construction activities.

Examples of the overall organisation structure of a construction firm and the organisation of a single project are given later in this chapter.

Commanding or directing

Commanding is the giving of instructions to ensure that the agreed policies are carried out. It also involves the granting of authority before commands can be issued.

Control

Control involves the continuous checking of performance with the plan and taking corrective action. The principles of control in relation to the control cycle are outlined in Chapter 6, and in Chapter 9 where control procedures have been applied to a project in relation to contract time (programme), money and cost (resources). Examples of control in construction include:

- Establishing budgetary control procedures within the organisation
- Control of cash flow (the movement of money)
- Control of project cost – monthly cost reporting
- Control of progress

Coordination

Coordination is the unification of effort between the company's personnel to ensure that the declared policies are fully implemented. Coordination is necessary between personnel in different departments to ensure the smooth running of the business.

Motivation

Motivation theories were developed by McGregor (1960) and Maslow (1954). Maslow's hygiene factors may be related to a modern contracting organisation by posing the question 'what motivates management personnel?'. Motivation factors include:

- Pay
- Job security
- Recognition, promotion
- The working environment and working conditions
- Office 'perks' such as company car, holidays, use of the company boat or flat
- Balance between work, pleasure and home life
- Creating the 'no blame' culture
- Company training culture and quality of training provision

Communicating

Communications are necessary to enable the business to function effectively. This involves the three-way transfer of information between client, contractor and staff.

Modern communication systems ensure that data can be transmitted rapidly by e-mail and fax, and electronic tendering procedures ensure that less paperwork is transferred between client and contractor at the bid stage of a project (Cartlidge (2002) explains how e-tendering works).

The office notice board is a useful means of communicating company policy statements, contract awards and staff promotions and appointments, and a company newsletter may help to boost morale and highlight regional activities and achievements.

2.2 Leadership styles

As businesses expand, the leadership style of the principal or managing director becomes apparent. The various leadership styles are summarised in Figure 2.1 as being autocratic or democratic.

Traditionally leadership has tended to be associated with autocratic command, especially within the small-sized organisation. Many still see leadership mainly in terms of issuing orders which are obeyed by subordinates without question. Drucker (1989) asserts that leadership is the lifting of man's vision to higher sights and the raising of man's performance to a higher standard. Management can only create the leadership under which potential leadership qualities become effective.

Clutterbuck and Goldsmith (1984) indicate that to be effective 'leaders must be seen'. Perhaps the question should be posed, 'How many times in the last twelve months have you had personal contact with your company chairman or chief executive?' Perhaps not at all?

Dixon (1991) defines leadership as the process of directing and influencing the work of team members. Leadership is concerned with guiding and directing others. The style adopted depends in part on the manager's view of human nature in general, and the ability of his/her subordinates in particular. The manager's attitude to his subordinates may be depicted by McGregor's Theory X and Theory Y, as outlined in Figure 2.2. The Theory X manager favours the autocratic approach while the Theory Y manager favours democracy.

Likert's system of management as outlined by Koontz and O'Donnell (1976) is an interesting approach which is worth considering. Likert

LEADERSHIP STYLES

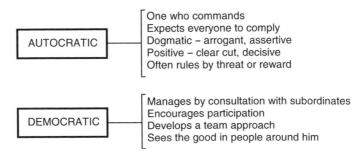

Figure 2.1

McGREGOR's THEORY X AND THEORY Y

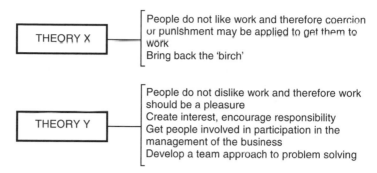

Figure 2.2

studied the pattern and style of leaders and managers for three decades in order to develop an approach to understanding leadership behaviour. He sees the effective manager as being strongly oriented to subordinates and relying on effective communications to keep all parties working as a unit. All members of the group, including the leader, adopt a supporting relationship in which they feel a genuine common interest in terms of needs, values, aspirations, goals and expectations. Likert sees this approach as the most effective way to lead a group.

Hodgetts (1982) presented a summary of the Likert system, as illustrated in Figure 2.3, which emphasises issues such as confidence and trust in subordinates and the process of decision-making. Likert found that those managers who applied the system 4 approach to their operations had the greatest success as leaders.

2.3 The size of construction firms

'Modernising Construction' (National Audit Office 2001) reported that there are over 160 000 firms operating in the construction industry and many of these are small contractors. In order to give a scale to the industry 'pyramid', the current annual turnover of the top 100 contractors in terms of annual turnover is shown in Table 2.1.

Table 2.1 represents less than 1% of the firms in the industry and indicates that the vast majority are either medium-sized or small. The annual turnover and profit of the top ten contractors is shown in Table 2.2.

Clearly, these firms are in their own league and it can be seen why the Major Contractors Group was powerful enough to dictate that the industry workforce must be certificated and competent under the Construction Skills Certification Scheme (CSCS) from the end of 2003.

LIKERT'S MANAGEMENT SYSTEM – Approach to Leadership

	SYSTEM 1	SYSTEM 2	SYSTEM 3	SYSTEM 4
Organisational variable	Explosive Authoritative	Benevolent Authoritative	Consultative Democratic	Participative Democratic
Leadership processes used	Have no confidence and trust in subordinates	Have condescending confidence and trust, such as master has to servant	Substantial but not complete confidence and trust, still wishes to keep control of decisions	Complete confidence and trust in all matters
Character of decision-making process	Bulk of decisions at top of organisation	Policy at top, many decisions within prescribed framework made at lower levels but usually checked with top before action is taken	Broad policy decisions at top, more specific decisions at lower levels	Decision-making widely done throughout organisation although well integrated through linking process provided by overlapping groups
Leadership style	AUTOCRATIC	AUTOCRATIC	DEMOCRATIC	DEMOCRATIC

Figure 2.3

Table 2.1 Top 100 contractors.

Turnover (£ million)	Number of firms
Less than 100	0
100–249	44
250–499	18
500–999	21
1000–1999	12
2000–2999	2
3000–3999	2
4000–4999	1

Table 2.2 Annual turnover and profit of top ten contractors.

Rank	Company	Turnover (£m)	Pre-tax profit (£m)	Profit (% of turnover)
1	AMEC	4331	39	0.9
2	Bovis Lend Lease	3755	67	1.8
3	Balfour Beatty	3441	88	2.6
4	Wimpey	2600	285	12.3
5	Taylor Woodrow	2215	233	11.7
6	Carillion	1974	42	2.1
7	Mowlem	1930	33	1.7
8	Barratt	1799	220	13.9
9	Persimmon	1711	256	17.5
10	Kier	1382	28	2.0

2.4 The characteristics of firms

It is interesting to compare the changes in the organisational structure of companies as company expansion or change takes place. In many construction situations the management are not able to cope with the management of changing size.

Drucker (1989) states that the biggest problem in business is growth, i.e. the problem of changing from one size to another. Many principals or owners of construction firms face this problem as business expansion takes place. Often they cannot cope with the new situation facing them. This is due to their lack of vision and competence to manage people around them. Many directors cannot delegate responsibility to subordinates due to a lack of trust and Drucker indicates that a change in behaviour, attitude, competence and vision is needed by people at the top.

Success in business often results from a company providing a good service to clients and doing a good job. The business can only service its customers by becoming bigger.

An interesting approach is taken by Drucker (1989) in respect of defining the four stages of business growth. These have been summarised as:

- The **small** business is distinguished from the one-man proprietorship by requiring a level of management between the man at the top and the workers. Also, small businesses tend to be organised functionally
- The **fair-sized** business (later referred to as medium-sized) – in this size of organisation, the role of the managing director has become a full time position. He is required to concentrate his efforts on guiding and managing the affairs of the business. The company needs some formal organisation structure which focuses the vision and efforts of managers directly on business performance and results. In the majority of construction firms in this category, the tendency is to develop a departmental approach to the various sections of the organisation, i.e. construction, surveying, estimating, plant, and administration
- The **large** business – at this stage of development the setting of overall objectives becomes far too big for one person and becomes a shared responsibility of the management team. A large construction organisation may be managed by a main board of directors, supported by departmental directors and regional and technical directors
- The **very large** business – this is characterised by the fact that the overall business objectives and resulting actions must be organised on a team basis. Each position in the organisation requires the full time services of several people

2.5 The small firm

National statistics categorise the small firm as a business with between one and twenty-four directly employed staff (or staff and operatives). This represents some 92% of companies in the UK construction industry. The number of construction firms in the European Union is in the order of 1 100 000 with 91% employing less than 10 people.

Business profile of small company A

Company background

- The business is managed by the principal or owner, with 60% of the work obtained by negotiation and the rest from competitive tenders. Quotations are based mainly on a drawings and specification basis, which places extensive risk on the business enterprise. Turnover

CASE STUDY – 'SMALL' COMPANY A

ORGANISATION STRUCTURE

RESPONSIBLE FOR:

PRINCIPAL OWNER

Contact with clients/architects etc.
Preparation of tenders/obtaining work
Submission and adjudication decisions
Pre-contract arrangements
Letting of subcontracts
Planning – programming of work
Interim certificates/final accounts
Payment to subcontractors/suppliers
Establishing policy
All decision making

Figure 2.4

PROJECT ORGANISATION

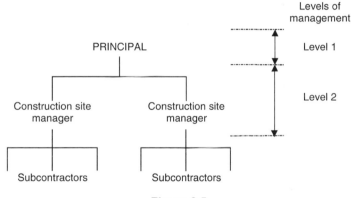

Figure 2.5

after 2 years of trading is in the order of £800 000. The largest project undertaken to date was in the £250 000 range

- The policy of the company is to use labour only and directly employed subcontractors. The number of directly employed staff is three. Figure 2.4 indicates the organisation structure of the company, if a structure can be said to exist

Figure 2.5 indicates the management approach to the control of projects.

Business control procedures

- Control is maintained by constant visits to projects by the principal in order to coordinate materials, plant and subcontractors. The principal's major concern is providing a personal service to clients and ensuring that projects are completed on time at a reasonable profit. A non-confrontational approach is adopted in the settlement of contract final accounts. The business relies extensively on negotiating further contracts with satisfied clients. Maintaining client contact during a project is considered essential to the success of the business

- Management checks are made on project profitability at interim valuation stages on a somewhat ad hoc basis. Monthly site meetings on all projects are attended by the principal. The majority of the principal's time is spent on preparing tenders for new work and chasing round his existing projects

Commentary on the role of the principal

- This is the typical small company approach to managing a construction enterprise. The principal tends to make all decisions, and tends to do everything himself. He does not appear to be able to delegate responsibility to people around him. With further business expansion the principal considers that he will lose control. Unless the principal overcomes this problem, the business will fail to expand and the principal will finish up having a heart attack!
- An extensive number of small companies in this situation may however be content with the profit returns they are achieving and may not wish to become any bigger. Perhaps the approach of working hard for nine months and spending 3 months lying in the sun is too much to give up for the sake of success!

2.6 The medium-sized firm

The medium-sized firm has developed from expansion of the small business, brought about by the increase in workload, turnover and business diversification – possibly away from the core business. This may have developed from a need for the company to consider broadening its work base. The core business may have been refurbishment work, but the company may now be offering a design and build or a work package service to clients. Alternatively, the business may have moved into its own speculative refurbishment projects. Many such options are available as a business expands.

A team approach to the management of the company has to be established by the appointment of directors responsible for contracts, surveying, estimating, office organisation and financial aspects of the business. As Drucker (1989) states, 'the role of the managing director now becomes a full time position'. A strategic plan and clear policy objectives must be established. Policy decisions must be communicated throughout the organisation, and the development of teamwork is necessary at all levels of management.

Policy must be clarified in such areas as:

- Safety, health and welfare
- Training and recruitment
- Company control and reporting procedures
- Planning procedures to be utilised on projects

Delegation of responsibility at all levels of management must be established. Control and reporting procedures must be established for reporting on contract performance during the progress of construction projects. Channels of communication must be established throughout the company in order that staff know the person they are responsible for reporting to.

Business profile of medium company B

Background information to company

- The company was established in the early 1980s as a two partner business undertaking public house and club refurbishment work. The company has now expanded into a major specialist contractor servicing the leisure and hotel industry. In 1995 the company turnover approximated to some £20 million. The company has recently expanded to offer a specialist design and build service to the leisure and hotel industry. The policy of the company is to use labour only and established domestic subcontractors. The company employs between 35 and 40 permanent office and site staff
- Figure 2.6 illustrates the current organisation structure showing the business divided into three functional areas, i.e. estimating/surveying, contracts, and office/commercial management. Figure 2.7 indicates the approach taken to the organisation of a major project. This simply consists of a construction manager who manages subcontractors. Head office support services are provided as indicated
- The company is in a very competitive market and pricing risks are taken due to the nature of their enquiries. In order to reduce the risk, they are moving towards offering a design and build service based on the client's scheme drawings or in-house design services. The largest single project undertaken to date was worth approximately £2–2.5 million

Business control procedures

- Control of the various functions of the business is delegated to a team of directors who cover construction management, estimating

CASE STUDY – 'MEDIUM' COMPANY B

ORGANISATION STRUCTURE

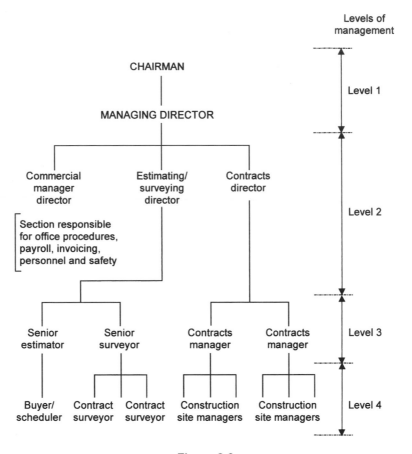

Figure 2.6

and surveying. The company is very surveying-orientated due to the extensive amount of subcontract work involved in the contracts. Cash flow is considered to be the life-blood of the company. Emphasis is on the control of time and money. Strict guide rules have been established for reporting on the cost and value situation at monthly intervals on all projects and an effective cost–value reconciliation system has been established. This provides data on the performance of each contract in the short term and provides the degree of control and reporting needed for overview by the directors

- Programming and progress reviewing is the responsibility of the contracts manager. Short-term planning and the coordination of subcontractors is the responsibility of the construction manager.

CASE STUDY – 'MEDIUM' COMPANY B

PROJECT ORGANISATION

RESPONSIBILITY LIES WITH CONTRACTS DIVISION

SITES ARE SERVICED BY HEAD OFFICE – BUYING/SURVEYING

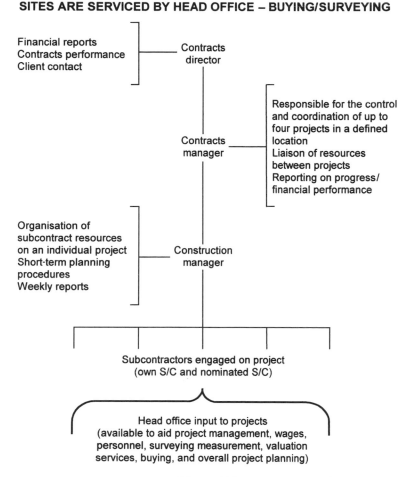

Figure 2.7

Strict control of subcontract orders is maintained by the link between the buyer and surveyors. Extensive pressure is applied to subcontractors to perform to programme. The success of projects relies largely upon subcontractor coordination and control

- Good client–contractor relationships are established early in the project by the appointment of a project director. This allows links between the client and contractor to be maintained at the top of the organisation and often leads to further negotiated contracts

2.7 A business unit approach in a medium–large firm

A medium–large north-west based contractor has taken an innovative approach to managing the business, involving establishing six business units within the organisation.

The business units are not based on the allocation of work within a region, but countrywide. Each business unit aims for a turnover of some £20–30 million each year. Certain units may be set up to service a single client (say on a partnering arrangement).

The aim of the business unit managers is to develop a team approach to managing projects that aids both company morale and clients alike. The role of the business manager is taken by a senior construction manager or it may be a shared position between a construction manager and senior quantity surveyor. Monthly reporting on progress and profit performance is made directly to the managing director and construction director.

The development of construction management project teams results in harmony between construction managers, site engineers, quantity surveyors and clients' representatives.

Figure 2.8 illustrates the application to the construction organisation.

2.8 The large firm

Most large contracting organisations develop from expansion of the medium-sized firm. The business may be managed from a single head office or may be regionalised in order to serve its customers' needs better. The functional approach to the management of the business will now have been expanded to form service departments, or divisions:

- Estimating
- Surveying
- Planning
- Contracts
- Plant
- Personnel
- Administration

Each may be managed by a director or, if regional, by a regional director. Figure 2.9 illustrates a large company divided into four company divisions responsible for construction activity. Figure 2.10 indicates the approach taken in the management of construction projects within the division. Head office is responsible for providing services to all projects in the form of office administration, estimating, surveying, project planning, safety and marketing.

BUSINESS UNIT APPROACH TO PROJECT MANAGEMENT

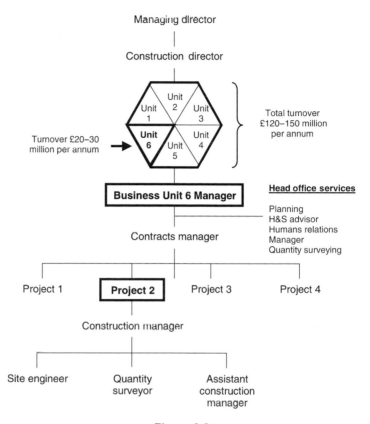

Figure 2.8

Figure 2.11 illustrates the overall site management structure for a £9.2 million management contract during the refurbishment of an inner city department store. The project involves both new work and refurbishment works while the store remains open for business. Figure 2.12 illustrates the organisation of the construction activity during the undertaking of the works on site.

Business profile of large company C

Background information to company

- The company was started in the early 1960s by the current chairman. In the early years, the objective was to develop a reputation as a quality housebuilder within the north-west region

LARGE COMPANY – CONSTRUCTION DIVISIONS

ORGANISATION STRUCTURE

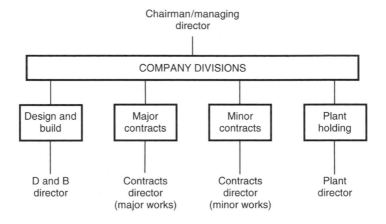

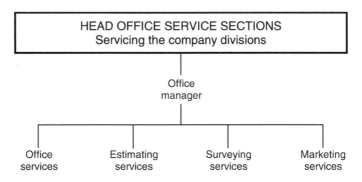

Figure 2.9

- By the mid 1970s this goal had been achieved, and in 1974 the company was incorporated as a private limited company. The business then diversified into small commercial office developments in order to provide a sound financial base for the future. At this stage a group holding company was established. Decisions regarding business diversification have led to further business success
- As further expansion took place, more ambitious commercial developments were taken on board. Development work now included shopping precincts, science parks and up-market commercial office projects. In the 1980s a range of villa developments were undertaken overseas in Spain, Portugal and the USA and in 1995 the company had a turnover of some £3 million from these projects. Further

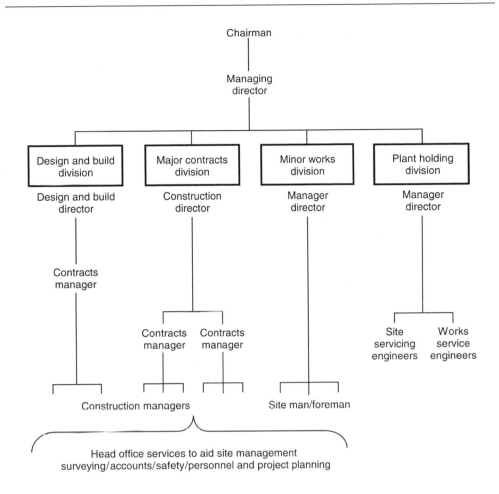

ORGANISATION OF CONSTRUCTION ACTIVITIES IN A LARGE COMPANY

Figure 2.10

diversification had also been made into the retirement homes market. Figure 2.13 illustrates the overall company structure and key figures from the 1995 accounts

- In 1995 the group turnover was approximately £57 million with profits in the region of £5 million. Development work was equally divided between its residential and commercial divisions. In the 1995 accounts, the company had fixed assets of £229 million and net assets of some £98 million; 99% of the share issue was held by the founder's family

- The group employed around 385 staff (including approximately 40 construction managers). The policy is to employ labour-only

SITE ORGANISATION – MAJOR PROJECT

OVERALL SITE MANAGEMENT STRUCTURE
(management project) – £9.2m value

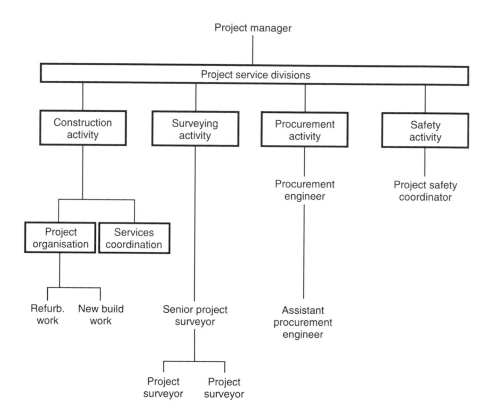

OVERALL ORGANISATION STRUCTURE

Figure 2.11

and domestic subcontractors, some of which have been with the company since the mid 1970s. As the group has grown, so have the subcontractors. Figure 2.14 illustrates the organisation structure of the residential homes division of the company's activities

- The homes division is currently undertaking a refurbishment project on a disused Victorian hospital. This involves refurbishing existing three- and four-storey blocks into luxury flats. Many Victorian features are being retained, including coved plasterwork to ceilings, shuttered windows and the recovery of pine doors and floors. The project value is approximately £10 million over a 3-year period. It is a credit to such companies to take an interest in restoring part of our

SITE ORGANISATION STRUCTURE FOR A REFURBISHMENT CONTRACT (Value £9.2m)

MANAGEMENT CONTRACT ARRANGEMENT

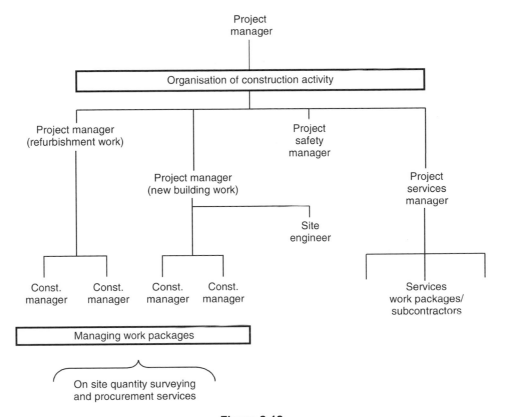

Figure 2.12

Victorian heritage by bringing such a building back into use for many people to enjoy

Business strategy and decision-making

- Decisions taken to diversify the range of business activities in the 1980s have certainly brought long-term financial returns
- Diversification into markets closely linked to the core business has proved successful. This has included such markets as retirement homes and commercial and industrial business parks, as well as overseas residential and villa development projects
- The company also supports a travel company to service their own holiday developments. Future proposals include establishing a joint

CASE STUDY – 'LARGE' COMPANY C

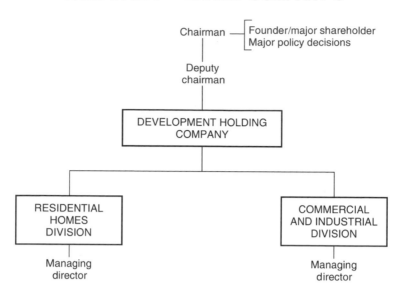

OVERALL COMPANY STRUCTURE

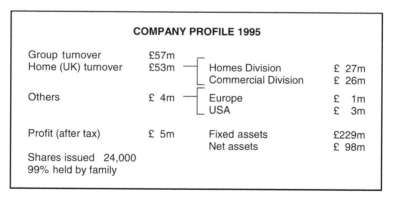

Figure 2.13

venture project with a major tour operator to build a villa holiday complex in the USA

2.9 Control procedures within organisations

As companies change size they need to review control procedures.

Denyer (1972), in his glossary of management terms, defines 'control' as the setting up of standards, making regular comparisons of actual events

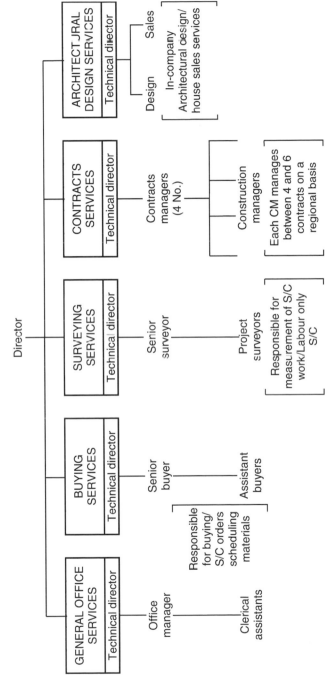

CASE STUDY – 'LARGE' COMPANY C

Director

GENERAL OFFICE SERVICES
Technical director

Office manager

Clerical assistants

Responsible for buying/ S/C orders scheduling materials

BUYING SERVICES
Technical director

Senior buyer

Assistant buyers

SURVEYING SERVICES
Technical director

Senior surveyor

Project surveyors

Responsible for measurement of S/C work/Labour only S/C

CONTRACTS SERVICES
Technical director

Contracts managers (4 No.)

Construction managers

Each CM manages between 4 and 6 contracts on a regional basis

ARCHITECT JRAL DESIGN SERVICES
Technical director

Design   Sales

In-company Architectural design/ house sales services

Total staff in Homes Division
Head quarters and site management
approximately 45 to 50 personnel

ORGANISATIONAL STRUCTURE RESIDENTIAL HOMES DIVISION

Figure 2.14

CONTROL PRINCIPLES

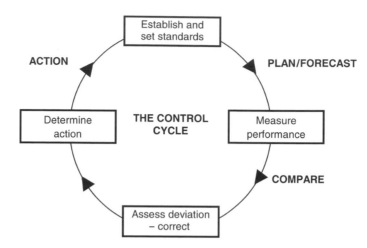

Figure 2.15

CONTROL APPLIED TO CHECKING DEVIATION FROM PROGRAMME

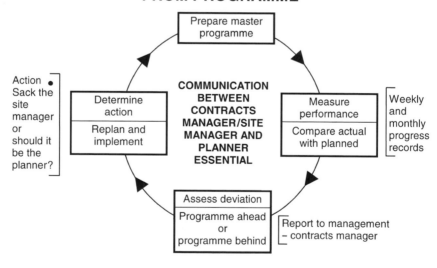

Figure 2.16

with those standards and taking corrective action. Figure 2.15 illustrates these principles diagrammatically in the form of a control cycle.

The application of the control process to progress reporting during a contract is illustrated in Figure 2.16. This illustrates the need for communication channels to be established between all levels of management. Denyer (1972) outlines various methods of control which may be applied to a construction situation, as follows.

Visual and personal

Checks applied by a supervisor or site manager. The regular visits to a project by the principal of a small organisation are a means of applying control to a project. The principal gets the feel that all is right or it quickly becomes apparent that all is not well – call it management intuition, if you like.

Reporting

The submission of regular reports during a project provides control

CONTROL IN ACTION

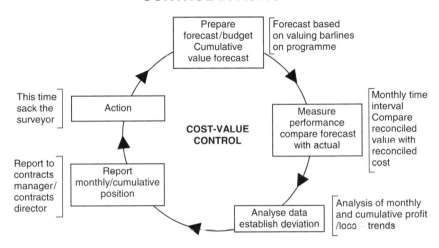

COST-VALUE PRINCIPLES

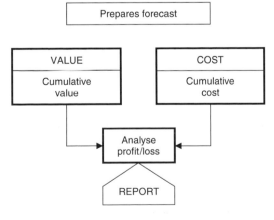

Figure 2.17

information to management. This applies to weekly site reports on progress and monthly cost–value reports on the profitability position. Monthly reports on progress and cash funding also aid management decision-making. As companies change size the value of internal reporting procedures needs to be considered.

Control by exception

This is a well-established management principle. It enables management to consider reporting only on serious deviations, otherwise it is assumed that performance is up to standard. Management by exception saves management time in analysing irrelevant reports and unnecessary data.

Control policy and the establishment of the necessary information in order to control the business must be established by senior management. The management must lay down policy statements in order to establish standards for controlling finance, time and quality. Control systems may be established for reporting on project performance and profitability. These will require strict monitoring and enforcement in order to succeed.

Figure 2.17 illustrates the principles of control applied to cost–value reconciliation procedures applicable to a construction situation. This is considered an essential control procedure in both the medium and large contracting organisation.

The application of project control procedures is dealt with in Chapter 9.

References

Cartlidge, D. (2002) *New Aspects of Quantity Surveying Practice*. Elsevier Butterworth-Heinemann.

Clutterbuck, D. & Crainer, S. (1990) *Makers of Management*. Macmillan Press.

Clutterbuck, D. & Goldsmith, W. (1984) *The Winning Streak*. Weidenfeld & Nicholson.

Denyer, J. C. (1972) *Students' Guide to Principles of Management*, 2nd edn. Zeus Press.

Dixon, R. (1991) *Management Theory & Practice*. Management Made Simple Series, Butterworth-Heinemann.

Drucker, P. F. (1989) *The Practice of Management*. Butterworth-Heinemann.

Hodgetts, R. M. (1982) *Management: Theory, Process and Practice*, 3rd edn. Holt-Saunders International Editions.

Koontz, H. & O'Donnell, C. (1976) *A Systems and Contingency Analysis of Managerial Functions*, 6th edn. McGraw Hill.

Maslow, A. H. (1954) *Motivation and Personality*. Harper.

McGregor, D. (1960) *The Human Side of the Enterprise*. McGraw-Hill.

National Audit Office (2001) *Modernising Construction*. Report by the Comptroller and Auditor General HC 87. The Stationery Office.

3 Managing risk

3.1 Risk in construction

Construction is undeniably a risky business for many reasons, including:

- Poor record of cost and time certainty for clients
- Adversarial attitudes and high levels of disputes and litigation
- The intense competition for work
- Low margins and profit risk
- The industry's poor safety and occupational health record
- Pressure from management and shareholders to produce a high return on funds invested
- Pressure on construction teams, especially site management and operatives, to save time and money
- Pressure on health and safety provision

Raftery (1994) argues that construction is nothing special compared with other industries because we simply carry out projects to a specified time-scale using teams of specialists for design and construction by marshalling appropriate resources to overcome the physical and technical problems involved. He also argues that construction projects exhibit much lower levels of technical complexity than aerospace, defence or computer software projects.

Smith (1999) distinguishes between risk and uncertainty in decision-making such that a risk is a decision having a range of possible outcomes to which a probability can be attached, whereas uncertainty exists if the probability of possible outcomes is not known. He suggests that risk falls into three categories:

(1) Known risks – risks that are an everyday feature of construction
(2) Known unknowns – risks which can be predicted or foreseen
(3) Unknown-unknowns – risks due to events whose cause and effect cannot be predicted

This point is also raised by Thompson and Perry (1992) and by Edwards (1995) who say that it is the unforeseen events which can have the most significant impact. Raftery goes on to judge that risk has to be recognised, assessed and managed and that overemphasis on risk avoidance leads to overcaution and negative attitudes.

3.2 The Turnbull Report

There are many risks in construction (in common with other industries) and this has been recognised by the Institute of Chartered Accountants who commissioned a report into the management of risk as an aspect of corporate governance. This is the Turnbull Report (ICA, 1999) and the recommendations from this took effect from the beginning of the year 2001.

Effectively, company directors have to show that they have proper and ongoing procedures to manage the risks to which their organisations are exposed and they have to demonstrate this in their annual report and accounts. This is not a mandatory requirement but is equivalent to other accounting standards and practices which accountants recognise. These standards are important to the extent that auditors, when they are not satisfied that the appropriate standards have been applied, will qualify the accounts accordingly. This is effectively a vote of no confidence in the accounts and can have a dramatic effect on the standing of the company and its ability to attract investors and customers. Compliance with the recommendations of the Turnbull Report has been a condition of Stock Exchange listing since January 2001.

Turnbull suggests that risk management is an 'ongoing process for identifying, evaluating and managing significant risks' and that this must be evident in the way the affairs of the business are conducted. The report identifies risk assessment as central to this process to the extent that 'the significant internal and external operational, financial, compliance and other risks' should be identified and assessed on an ongoing basis. These significant risks include:

- Market
- Credit
- Liquidity
- Technological
- Legal
- Health, safety and environmental
- Reputation
- Business probity

3.3 Project risk

Edwards (1995) suggests that identifying hazards is an essential part of a structured approach to risk management but that it is often unidentified hazards, for which no provision has been made, that have the most significant impact. Thompson and Perry (1992) observe that all too often risk is either ignored or dealt with in an arbitrary way on construction projects, and that the practice of adding a 10% contingency is typical industry practice. Alternatively, the contractor may 'take a view' to cover for any shortfall in the estimate or inaccuracy in the client's documentation.

Figure 3.1 illustrates that risk assessment will not remove all risks on a project and that both the client and contractor must recognise that residual risks will always remain no matter what provisions are made. The aim of risk management is to ensure that such risks are managed effectively by the party best able to do so.

RISK MANAGEMENT PROCESS

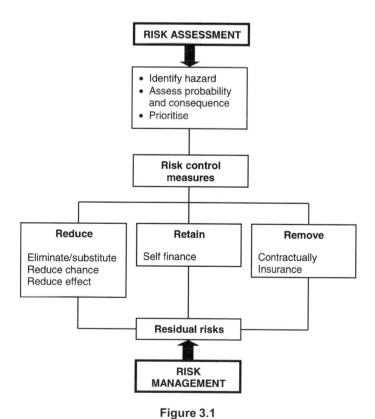

Figure 3.1

3.4 Risk management

The extent to which the risk can be reduced or controlled is the responsibility of management, whose function is to plan, organise, control, monitor and review the measures needed to prevent exposure to risk. To do this requires identification of the hazard, an assessment of the extent of the risk, the provision of measures to control the risk and the management of any residual risk remaining.

Unfortunately, despite all the steps that designers and managers can take, humans and human systems are imperfect and accidents and losses can and do happen.

Therefore, a roofer working at height has a chance of falling and suffering serious or fatal injury unless:

- The work can be avoided (i.e. designed out), or
- Suitable measures can be installed to prevent him falling, and
- Suitable safeguards can be provided should he happen to fall in any event

One common method of managing risks is to devise a risk register for the project. An example is shown in Figure 3.2. The CIOB Code of Practice for Project Management (CIOB 2002) defines this as a 'formal record for risk identification, assessment and control actions' which may be divided into three parts:

- Generic risks – risks which are present irrespective of the project type or nature

RISK REGISTER

Item	Description	Risk			Without controls			Controls	Residual risk	Action
		H	M	L	Cost impact	Time impact	Other			

Figure 3.2

- Specific risks – risks which are particular to the project in hand
- Residual risks – remaining risks despite the control measures proposed

Thompson and Perry (1992) suggest that risks that have not been allowed for lead to the concept of risk exposure.

3.5 Risk assessment

Management clearly needs to be concerned about these issues, both in the client organisation and within the contractor. The bottom line is the prevention of losses which are caused by exposure to risk. These losses include:

- Accidents to workers
- Accidents to the general public
- Loss of skill and experience
- Damage to property
- Loss of time and production
- Loss of money on contracts
- Loss of reputation and future business

Where there is no risk there is no chance of a loss occurring, but in the real world risk is all around. Risk is caused by hazards. BS 4778 defines a hazard as 'a situation that could occur that has the potential to cause human injury, damage to property, damage to the environment or economic loss'.

In order to control risk, a risk assessment is carried out. This is a statutory requirement under the management of Health and Safety at Work Regulations 1999 with regard to risks to health and safety. The risk assessment might follow the steps listed in Table 3.1.

Risk is measured by evaluating the chance of something happening (the hazard) and the severity or consequences if it does. This can be based on a hunch or by making an intuitive assessment of whether the risk is high, medium or low or by using a matrix to do a simple calculation (Table 3.2).

Therefore, on a scale of 1–3, if the likelihood (chance) is 2 and the severity is 3, the risk is 6 out of 9. This would be in the high risk category. However, it must be remembered that the prediction of likelihood and severity is difficult to do with any degree of certainty.

The matrix could be used to assess commercial risk at tender stage or the risk due to a variety of possible ground conditions or the chance of a fatal or non-fatal accident to people. The effect of exposure to this risk might be the loss of money, loss or damage to property, or business losses including loss of reputation or loss of individual skill and experience due to an accident to a key worker.

Table 3.1 Risk assessment.

Step	Example
1 Identify the hazard	Deep drainage excavation in bad ground
2 Identify who or what might be harmed	Pipelayers in trench
3 Evaluate the risks arising from the hazard	High risk of collapse
4 Determine the control measures required	Use steel trench drag box
5 Evaluate remaining risks	Risk of crushing/injury from excavator bucket Risk of falling materials
6 Record the findings of the risk assessment	Fill in risk assessment sheet
7 Make contingency plans for the residual risks	Prepare safety method statement based on risk assessment Supervisor to give task talk Permit to work required Banksman working with excavator
8 Review and revision	Monitor site operations and modify risk assessment where necessary Hold further task talk if method statement is changed

Table 3.2 Risk matrix.

Severity	Likelihood		
	1	2	3
1	Low	Low	Medium
2	Low	Medium	High
3	Medium	High	High

3.6 Client risk

Thompson and Perry (1992) suggest that uncertainty is greatest at the earliest stages of a project and that time and cost overruns can invalidate the client's business case for a project by turning a potentially profitable venture into a loss-maker. They also emphasise that the risks with the most serious effects for clients are:

- Failure to keep within the cost estimate
- Failure to achieve the required completion date
- Failure to achieve the desired quality and functional requirements

PROCUREMENT RISK

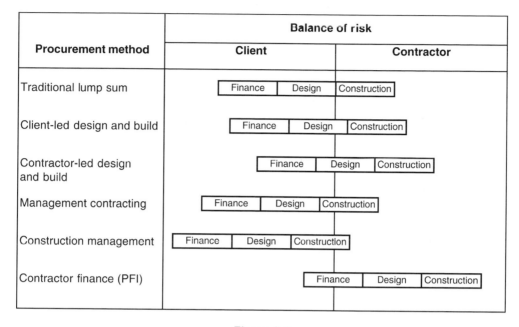

Figure 3.3

Thompson and Perry's work emphasises that appropriate strategies are necessary for the control and allocation of risk and that, while risk cannot be eliminated through procurement, contractual arrangements can greatly influence how risks are managed. This is illustrated in Figure 3.3 which provides a simplified view of the balance of risk according to the procurement strategy chosen. This diagram is based on the work of Walker and Greenwood (2002) and Edwards (1995).

Some of the risk management considerations made by a client at project level are considered here.

Feasibility risk

At the early stages of projects clients must confirm the business case, identify options and develop the preferred solution. Once a scheme is sanctioned by the client, major commitments are made in terms of design, procurement and construction.

Design risk

Business decisions are all about risk and reward and the client must decide how much control is required over the design of the project.

Retaining control over the quality of the design may be important for the client but the downside is that the design risk remains with the client as well. Partial or complete contractor design may remove the risk but then the client risks getting the wrong end result.

Funding risk

The contractor's income is the client's negative cash flow and arrangements must be made for available funds to draw down in order for the client to make regular monthly or stage payments for work in progress. The client's quantity surveyor must be careful not to expose the client to the risk of the contractor's insolvency by overvaluing interim payments.

Tender documentation risk

The traditional bills of quantities contract came about in order to give tendering contractors a level playing field. If traditional procurement is used and provided the design is well developed before going out to tender, quantities will be firm and all other documentation complete. Clients risk contract variations and budget overspend if they push for a fast start on site, and they may be well advised to consider more appropriate procurement arrangements in such cases.

Time risk

The obligation to complete the project on time is the contractor's responsibility and the client has redress in standard contracts through the liquidated and ascertained damages (LAD) provisions. Despite being a genuine estimate of the client's loss, LADs may not be adequate compensation if the contractor is late finishing, especially if a sell-on deal is lost or a prestigious opening deadline is missed.

Commercial risk

For most clients, buildings represent assets which are used to generate income and profits. The commercial success of a project may well be undermined if the job is delivered late or over budget or if the quality of design and construction is below the necessary standard. Clients should engage competent professional representatives to make sure this does not happen irrespective of the procurement method used for the project.

3.7 Contractor risk

Some of the biggest risks taken by contractors are at tender stage when they commit to a price and programme. Many companies now consider risk management to be an essential part of the tendering process. A contractor's risk assessment at the estimating stage may include consideration of the following risk areas.

Tender risk

At tender stage the contractor needs to consider many factors before submitting a bid. Among these are:

- Previous experience (good/bad) working with the client team
- The financial stability of the client
- Market conditions and the level of competition for the contract
- Inflation – is a firm or fluctuating price required?
- Ground conditions and the balance of risk in the contract and method of measurement

Risks are involved with checking each section of the estimate build-up to ensure that the estimator's allowances are realistic. Final adjustments, the tactical movement of moneys in the bills of quantities and the distribution of overheads and profit among the trade sections can expose the contractor to risk once the contract is awarded.

Quantity risk

The contractor must assess the accuracy of the quantities in the bills at tender stage because margins can be lost if the quantity work is subsequently reduced on remeasure. Provisional quantities in the bills pose a risk in rating up and allocation of margin.

Where quantities have been taken off by the contractor, failure to include associated labour items may add risk to the overall bid. On design and build projects, the contractor is responsible for taking off his own quantities but many serious quantity errors are often not apparent until the contract has been awarded. The contractor must then simply accept the error and grin and bear the loss (or find another way of making up the money).

Subcontractor risk

On many contracts, the contractor may simply be responsible for managing subcontractors, with very little work directly under his control. The ultimate success of the project may lie in the performance and organisation of subcontract operations. The contractor takes all the tendering risk and is especially vulnerable to subcontractors who are unwilling to stand by their quotes at tender stage.

The following check procedures may be implemented and when undertaken in conjunction with a well-managed workload should reduce the risk to the company:

- Subcontract totals should be abstracted from priced bills and checked against individual trade sheets
- The correct selection of subcontractors and suppliers to ensure suitability for the work must be considered
- Ensure liaison with construction, surveying and the buying department to ensure that the best subcontract packages have been selected
- Subcontract price comparisons should highlight potential high/low rates so that assessment may be made in tender settlement meetings
- The design input of work packages should be compared to ensure that the quotations meet design requirements. Likely cost savings proposed by subcontractors in relation to time and quality should be highlighted. Variations from the norm should be investigated

Design risk

The contractor may be responsible for temporary works design only or may be involved in partial or complete design of the permanent works. On design and build contracts (JCT 98 with Contractor's Design) or on JCT 98 with Contractor's Designed Portion Supplements, the risk in each design package must be carefully assessed. The contractor is responsible for the quality of the work of subcontractors. Lead-in times for design packages must be carefully considered at tender stage so that the contractor is not committed to an unrealistic programme.

Programme–time risk

Where the time for completion is stated in the tender documents, the contractor may be at risk if the client/project manager has got it wrong. On the other hand, where the tender documents require the contractor to insert his own assessment of the contract period, the contractor will be gambling on his own judgement.

It is essential for the success of the project that a realistic assessment of the construction period is backed up by a comprehensive pre-tender programme. The build-up of the contract preliminaries is based on the programme, together with major items of plant. Including contingencies for liquidated damages for non-completion (where a tight programme period has been stated) only makes the tender less competitive.

Method risk

The contractor's choice of construction method at the tender stage is crucial to winning the contract, but also fraught with risk. The ground conditions on site may be different to those expected and the type of earthwork support required may be more expensive than that allowed for in the tender. Relief may be obtained through the method of measurement but this is not always the case, especially on civil engineering projects.

It may be that the estimator's choice of plant or cranage is not feasible once the contract has been awarded, but the contractor is stuck with the allowances in the tender. This is where the contracts manager earns his money because cost effective ideas will be needed to prevent contract losses.

Health and safety risk

Health and safety risk arises from the impact of hazards. Where there is no hazard there is no risk, but in construction there are hazards everywhere on a site. The best that can be done is to eliminate hazards in the design of the building and reduce the possible effects of residual risks through good management. The effective planning, organisation and control of construction work is central to that process.

People at risk from construction work include:

- Persons at work
- Visitors to site (including client representatives)
- General public
- Children

Documentation risk

Clarity of tender documentation is important. Bills of quantities containing extensive provisional quantities need careful pricing. Prices based on drawings and specification, or schedules of work containing extensive spot items, may prove difficult to price accurately.

Where there are defined provisional sums in the bills of quantities, the contractor has to allow for this work in his programme, even though there may be little information as to the precise nature of the work required.

The contractor needs to scrutinise the tender documentation very carefully in order to assess the implications of:

- Onerous contract terms
- Clauses deleted from standard contracts
- High levels of liquidated damages
- Unrealistic contract period
- Possible novation of the design (design and build contracts)
- Contract bonds and guarantees required

After consideration of all the risks to be assessed at the tender stage, it may be better for the contractor to withdraw from the competitive tender altogether, and spend his time negotiating work (or even consider partnering).

3.8 Tendering risk

When competition for work is fierce, contractors have to find ways of winning work *and* making a profit at the end of the job. The starting point is the estimate produced by the estimator.

Many contractors will carry out a risk assessment at tender stage and this may identify areas for making savings in the tender figure. The contractor may spot undermeasure or overmeasure in the contract bills or there may be scope for variations in the documents. Alternatively, the contractor may take into account buying 'muscle' on suppliers' and subcontractors' prices as a means of reducing the tender figure. This is frequently referred to as commercial opportunity or scope.

In the example in Table 3.3, it can be seen that the contractor's adjustments bring the tender figure (£1 780 500) below the estimator's net cost (£1 800 000). Effectively, the contractor is tendering at below net cost or, in other words, tendering at a negative margin. This is achieved

Table 3.3 Tender summary.

		£	£	£
Preliminaries			130 000	
Measured work	Labour	100 000		
	Plant	60 000		
	Materials	240 000		
	Subcontract	820 000	1 220 000	
PC sums			300 000	
Prov. sums			150 000	
Net total				1 800 000
Overheads and profit 7%				126 000
Total				1 926 000
Commercial opportunity:				
Materials	7.5%	240 000	18 000	
Subcontract	10%	820 000	82 000	
PC sums	2.5%	300 000	7 500	
Provisional sums[1]	8%	150 000	12 000	
Preliminaries[2]	4 weeks	6 500	26 000	(145 500)
Tender total				**1 780 500**

1 Assuming rates for this work will include 7% mark-up and allowing for commercial opportunity of, say, 5% on materials and subcontractors

2 Assuming a 4-week reduction on the 20-week contract period and that all preliminaries costs are time-related

simply by transferring the risk to others, principally the domestic subcontractors.

The contractor is taking a gamble in that he might not be able to squeeze down subcontractors' prices once the contract has been awarded, or the anticipated returns from variations and claims may not be forthcoming. This does not mean that the contractor will necessarily lose money on the contract; all will be well provided that allowances in the tender can be achieved in reality.

The tender documentation offers the contractor many opportunities to take commercial advantage, which can be used as a means of winning contracts and making money during the contract period. For instance, documents are frequently littered with mistakes or inaccuracies and contractors can take advantage of these. In such circumstances, the contractor might spot that the concrete work is undermeasured and that the architect will have to issue variation instructions for additional work.

At tender stage the contractor can set the scene to make money out of this situation during the contract. By increasing his bill rates for these items of work, which will subsequently be remeasured and valued during the contract, the contractor can 'load' the rates and thus make more profit. To do this without increasing the tender sum, the contractor will take money away from some bill items and reallocate it to the undermeasured items. This technique is sometimes called rate loading.

Tender loading is a similar technique and is another way for the contractor to move risk on to the employer and make money at the same time.

Front-end loading reduces the early negative cash flow effect by increasing the margin. This is done by pricing the bill of quantities so that the margin is allocated to those items which will be carried out during the early stages of the project.

Back-end unloading is a similar technique but involves increasing both margin and net cost on early items of work. This is a dangerous practice and involves moving part of the net cost allowance for later items of work in the bill of quantities to those which are to be carried out earlier in the contract.

Harris and McCaffer (2000) explain these techniques further.

3.9 Health and safety risk

On average two people die every week on construction sites but studies have shown that 90% of these deaths could be avoided and 70% of these lives could have been saved by positive management.

Construction health and safety risk is managed through legislation and in particular:

- The Health and Safety at Work Act 1974
- The Management of Health and Safety at Work Regulations 1999
- The Construction (Design and Management) Regulations 1994
- The Construction (Health, Safety and Welfare) Regulations 1996

Other legislation, which is not construction industry specific, also applies to construction projects.

The Health and Safety at Work Act 1974 is enabling legislation which sets out the penalties for failure to comply, and confers powers on inspectors who may require site practices to be changed if they are considered unsafe or in breach of legislation.

A contractor can be served with an improvement notice or may even be prevented from continuing an unsafe operation through a prohibition notice.

The Management of Health and Safety at Work Regulations 1999 deal with assessment of risk and arrangements for and competence in the measures needed to protect individuals and prevent accidents at work. The impact of this legislation is to ensure that issues identified by risk assessments are dealt with by effective planning, organisation and control and that procedures to monitor and review such arrangements are put in place.

The Construction (Design and Management) Regulations 1994 require a health and safety plan for all but the smallest of projects and this acts as an important means of communication which is intended as a continuous theme through the project.

In practice, the health and safety plan has two parts:

- Pre-tender health and safety plan
- Construction phase health and safety plan

The pre-tender plan is prepared by the design team in conjunction with the planning supervisor, while the construction phase plan is developed from this by the principal contractor into a safety management system for organising, implementing and monitoring health and safety arrangements on site.

The Approved Code of Practice (HSE 2001) gives comprehensive guidance concerning the contents of the health and safety plan and explains the responsibilities of those who have duties under the CDM Regulations:

- Client
- Designers
- Planning supervisor (statutory appointment)
- Principal contractor (statutory appointment)
- Contractors

The principal contractor role is normally undertaken by the main contractor under the contract, although the statutory appointment and the appointment of the contractor under the civil contract are quite separate and distinct appointments. The duties of the principal contractor include responsibility for the health and safety management system for the project. This normally involves conducting site inductions, establishing site rules, the provision of training and information and the prevention of unauthorised access to the site.

3.10 Fire risk

Fire is an ever-present risk on construction sites, especially with respect to:

- Hot work such as welding, blowlamps, cutting and grinding
- Heating appliances, especially gas bottles in welfare facilities
- Litter, especially in rest rooms and drying areas
- Arson
- Smoking
- Burning of waste on site
- Stored materials, including adhesives and solvents

The Construction (Health, Safety and Welfare) Regulations 1996 provide for such eventualities and require contractors to:

- Take measures to prevent risk of injury from fire
- Provide and maintain fire-fighting equipment, fire detectors and alarm systems
- Provide access for fire-fighting equipment
- Give instructions to people in the use of fire-fighting equipment
- Give instructions to people where their work activities involve a fire risk
- Indicate fire-fighting equipment with suitable signs

Standard forms of contract provide for insurance of the works during construction and this includes the risk of fire. However, the high cost of site fires has prompted the publication of a Joint Code of Practice, 'Fire Prevention on Construction Sites' (BEC, 1995). This provides that potential fire risks are considered at design stage and also that a site fire safety plan is developed by the principal contractor. This plan must include provisions for a comprehensive fire prevention regime including:

- Organisation and responsibilities for fire safety
- Hot work provisions

- Fire escape and communications
- Fire drills and training
- Emergency procedures

The fire safety plan may be included in the construction phase health and safety plan. It should be noted that adherence to the Joint Fire Code is a contract condition in some standard forms (e.g. JCT 98).

References

BEC (1995) *Fire Prevention on Construction Sites,* Joint Code of Practice. Construction Confederation.

CIOB (2002) *Chartered Institute of Building Code of Practice for Project Management,* 3rd edn. Blackwell Publishing.

Edwards, L. (1995) *Practical Risk Management in the Construction Industry.* Thomas Telford.

Harris, F. & McCaffer, R. (2000) *Modern Construction Management,* 5th edn. Blackwell Publishing.

HSE (2001) *Managing Construction for Health and Safety,* Approved Code of Practice and Guidance. HSE Books.

ICA (1999) *The Turnbull Working Party Report.* Institute of Chartered Accountants.

Raftery, J. (1994) *Risk Analysis in Project Management.* E. & F.N. Spon.

Smith, N. J. (1999) *Managing Risk in Construction Projects.* Blackwell Science.

Thompson, P. & Perry, J. (eds) (1992) *Engineering Construction Risks.* Thomas Telford.

Walker, P. & Greenwood, D. (2002) *Risk and Value Management.* RIBA Enterprises.

4 Programming techniques

4.1 Introduction

Without planning it is difficult to envisage the successful conclusion of any project or the effective control of time, money or resources. Planning is also essential in order to deal with construction risks and devise safe working methods. This is true throughout all stages of the process from inception through the design, tendering, construction and commissioning stages of a project.

The reasons for planning may be summarised as:

- To aid contract control
- To establish realistic standards
- To monitor performance in terms of output, time and money
- To keep the plan under constant review and take action when necessary to correct the situation

Chapter 9 discusses control as an essential requirement of a good planning strategy.

A variety of programming techniques are available to the client's project manager or the contractor's planner and these can be used according to the type and complexity of the project concerned. The following will be considered in this chapter:

- Bar charts
- Linked bar charts
- Arrow diagrams
- Precedence diagrams
- Line of balance
- Time-chainage diagrams

The technique adopted in any particular case is largely a matter of personal preference but the recipient of the information needs to be considered as programmes based on overly-complex techniques may be counter-productive to effective communication.

Table 4.1 Planning programmes.

Planning stage	Type of programme
Design	Project master schedule
Tender	Pre-tender programme
Pre-contract	Master programme
	Target programme
	Subcontractors' programme
	Procurement programme
Contract	Stage programme
	Short-term programme
	As-built programme

Programmes are commonly prepared at various stages of projects for different purposes and different audiences and these are summarised in Table 4.1.

4.2 The planning process

There is more to planning a project than meets the eye and a great deal more involved than simply producing a programme. Both the client's representative or project manager and the contractor will have many issues to think about if the project is to be successfully completed.

Figure 4.1 summarises some of the many complex considerations made during the client's planning process, and Figure 4.2 shows a similar overview from the contractor's viewpoint. Figure 4.2 establishes the relationship between the three planning stages undertaken by the contractor and a check-list approach is suggested for each stage to illustrate what needs to be considered. The considerations suggested are those generally undertaken by both medium-sized and large contractors but are equally valid for small contractors as well.

Clearly there will be an interface between the client's planning process and that of the contractor at the tender stage of a traditional project, and during the administration of the contract on site. Where non-traditional procurement methods are used, there may be further interfaces to consider.

Planning is one of Henri Fayol's six functions of management and it starts right at the outset of a project. There are several levels of planning:

- Project planning carried out by the client/project manager
- Pre-tender planning carried out by the tendering contractors
- Pre-contract planning carried out by the main contractor
- Contract planning carried out by the main contractor and subcontractors

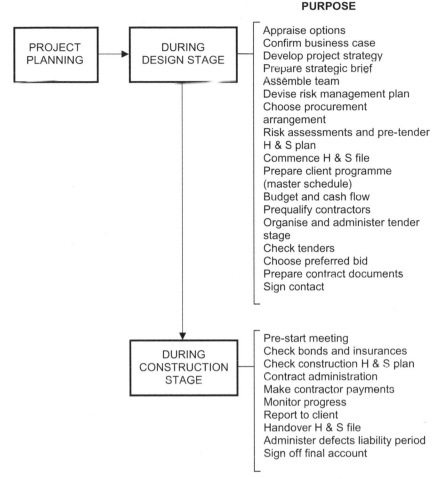

PURPOSE

PROJECT PLANNING → DURING DESIGN STAGE

Appraise options
Confirm business case
Develop project strategy
Prepare strategic brief
Assemble team
Devise risk management plan
Choose procurement arrangement
Risk assessments and pre-tender H & S plan
Commence H & S file
Prepare client programme (master schedule)
Budget and cash flow
Prequalify contractors
Organise and administer tender stage
Check tenders
Choose preferred bid
Prepare contract documents
Sign contact

DURING CONSTRUCTION STAGE

Pre-start meeting
Check bonds and insurances
Check construction H & S plan
Contract administration
Make contractor payments
Monitor progress
Report to client
Handover H & S file
Administer defects liability period
Sign off final account

OVERVIEW OF CLIENT PLANNING PROCESS

Figure 4.1

4.3 Project planning

Project planning starts with the client team or organisation when a programme (or master schedule) will be prepared by the client's agent, representative or project manager. This sets out the broad framework for the project, including:

- Key dates for commencement and completion of project
- Key dates for design, tender and construction
- Overall programme and phasing
- Design and tendering periods
- Key dates for commencement and completion of construction
- Stage or phased handover dates

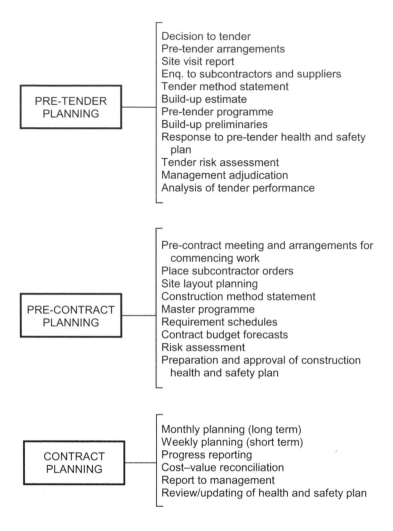

OVERVIEW OF CONTRACTOR PLANNING PROCESS

Figure 4.2

Reasons for project planning

- To establish a realistic project master schedule or programme on which to base the appointments of consultants and contractors and commission the occupancy phase of the scheme
- To identify key dates or gateways at critical stages of the project
- To facilitate control of the design and tendering process
- To identify potential risks to progress and avoid possible delays to project completion and revenue generation

- To facilitate the arrangement and draw-down of client cash funding in order to pay for design and other professional services and make interim payments to contractors
- To establish a realistic time period for the construction stage
- To monitor actual progress and take corrective action

4.4 Pre-tender planning

At the tender stage, a pre-tender programme is usually prepared by the contractor as an aid to the tendering process. The pre-tender programme will assist the estimator to price key method-related items in the bills of quantities and also the contract preliminaries which comprise largely time-related costs.

Pre-tender planning may be defined as the contractor's planning considerations during the preparation of an estimate and its conversion into a commercial bid. The role of pre-tender planning is shown in Figure 4.2 which provides an overview of the planning process from the contractor's point of view. This clearly illustrates that pre-tender planning involves much more than simply producing a programme for tendering purposes, but concerns all aspects of a project from the initial enquiry from the client to the submission of the contractor's tender bid.

Reasons for pre-tender planning

- To establish a realistic contract period on which the tender may be based
- To identify construction methods
- To assess method-related items which affect the bid price
- To aid the build-up of contract preliminaries and plant expenditure
- To aid the tendering process

4.5 Pre-contract planning

Pre-contract planning generally takes place during the period between contract award and commencement of work on site. This is the case for a project based on a traditional competitive tender but there may be differences in procedures where other procurement arrangements are used.

Before work starts on site, the contractor will develop the pre-tender programme into the contract master programme showing the main construction operations to be carried out. Copies of this programme will be presented to the client's representative who will use it as a tool to monitor the contractor's overall progress during construction. The master programme will often show when information is required by the contractor and act as a prompt for the architect.

The master programme is the one that the client team sees, but many contractors produce an internal programme for their own use in order to save time and money. This is called a target programme and is effectively a compressed version of the master programme with time taken out of the critical path. This is a commercial decision which may not work out as planned – that's contracting!

In order to help the contractor organise and manage site activities at an operational level, the target programme will need to be developed in more detail. Most of the time bars on the target programme will represent the main work packages to be carried out by various subcontractors and consequently each bar will be developed into a subcontractor programme showing the detailed activities to be carried out.

However, to ensure that work packages start and finish on time, it is now usual for the contractor to produce a procurement programme for each subcontractor. This programme will show both negative time and positive time. Negative time is the time needed before work starts on site to organise design and fabrication aspects of the package or to pre-order key materials with long lead times. Positive time is the time needed to carry out the subcontractor's work on site.

Reasons for pre-contract planning

- To provide a broad outline plan or strategy for the project
- To comply with contract conditions
- To establish a construction sequence on which the master programme may be based
- To identify key project dates
- To highlight key information requirements
- To enable the assessment of contract budgets and cumulative value forecasts
- To schedule key dates with respect to key material and subcontractor requirements

4.6 Contract planning

During the contract stage, the master programme will be further developed. For instance, a stage programme might be prepared showing part of the master programme in more detail. Alternatively, the contractor might produce a series of short-term programmes at weekly or fortnightly intervals so as to plan day-to-day work in detail.

Contract planning is done by the main contractor in order to maintain control and ensure that the project is completed on time and within the cost limits established at the tender stage. Subcontractors contribute to the process either by submitting their work programme for approval or through discussion with the main contractor.

As the contract progresses, invariably the programme changes from its original form. Delays occur, work is disrupted due to design changes and unforeseen events take place such as the discovery of bad ground or contamination. This causes delay and/or disruption to the programme which the contractor has to accommodate. These changes should be recorded on a revised programme which should be constantly updated throughout the project as work proceeds and as other problems arise. These programmes are often referred to as the as-built programme or, alternatively, the programme of the day and they are a vital tool to enable the contractor to justify his entitlement to extensions of time and/or additional payment for loss and expense.

Reasons for contract planning

- To monitor the master programme – monthly, weekly and daily
- To plan site operations in detail in the short term
- To optimise and review resources
- To keep the project under review and report on variances

4.7 Method statements

The preparation of method statements forms an essential part of the contractor's planning process as these underpin the programme and explain how the work is to be undertaken. Method statements may be categorised into three distinct formats:

- The tender method statement
- The construction or work method statement
- The safety method statement

Figure 4.3 summarises the purpose of each type of method statement, which may be presented in a written or tabular form.

The tabular format is probably simpler and easier to read but, as with all standard forms, there is a space restriction on the content. Using a written or prose format (rather like a mini report) gets over this problem and, provided it is well laid out, can be both useful and comprehensive. Both types are commonly used, according to personal preference. Whichever is chosen they should not be too long or overly complex as

PURPOSE

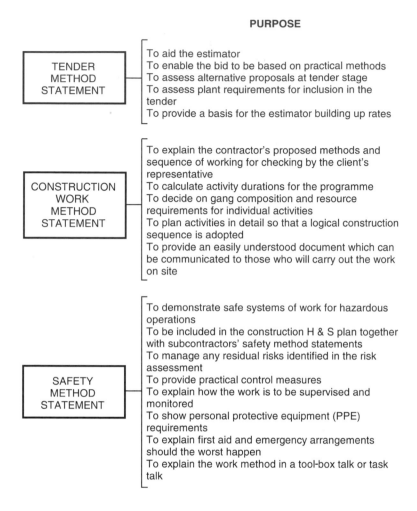

| TENDER METHOD STATEMENT | To aid the estimator
To enable the bid to be based on practical methods
To assess alternative proposals at tender stage
To assess plant requirements for inclusion in the tender
To provide a basis for the estimator building up rates |

| CONSTRUCTION WORK METHOD STATEMENT | To explain the contractor's proposed methods and sequence of working for checking by the client's representative
To calculate activity durations for the programme
To decide on gang composition and resource requirements for individual activities
To plan activities in detail so that a logical construction sequence is adopted
To provide an easily understood document which can be communicated to those who will carry out the work on site |

| SAFETY METHOD STATEMENT | To demonstrate safe systems of work for hazardous operations
To be included in the construction H & S plan together with subcontractors' safety method statements
To manage any residual risks identified in the risk assessment
To provide practical control measures
To explain how the work is to be supervised and monitored
To show personal protective equipment (PPE) requirements
To explain first aid and emergency arrangements should the worst happen
To explain the work method in a tool-box talk or task talk |

OVERVIEW OF METHOD STATEMENT FORMATS

Figure 4.3

they are an essential aid to conducting activities on site in accordance with a well thought-out and agreed procedure, and they need to be easily communicated at site level.

Illingworth (1993) gives examples of the formats used in practice. It is essential that the purposes of method statements at the various stages of the planning process are understood. Method statements convey different meanings to clients, consultants, subcontractors and main contractors. A method statement is not simply a list of construction operations with notes written alongside.

It is common practice to combine the construction and safety method statement for each work activity into a single document. This makes sense because safety is an integral factor in the planning process. A disadvantage of doing this is that the combined method statement may become too complicated.

Further explanations and worked examples of method statements are given in Chapters 5, 8 and 10.

4.8 Planning a project

When preparing any programme for a project it is essential to follow a logical thought process in order to develop a realistic and workable programme. A working knowledge of the development and construction processes is essential. The level of detail shown in the programme should be commensurate with the project stage under consideration and, wherever possible, activity durations should be based on empirical data or calculation. Computer software should be used for speed, for considering 'what if' options and for high quality professional presentation.

The planning of a project requires a logical approach involving various steps or thought processes:

- Getting a feel for the project
- Establishing key project dates
- Establishing key activities or events
- Assessing how long the activities will take
- Establishing the sequence
- Deciding which programming technique to use

These are now considered in turn.

Getting a feel for the project

It is important for the planner or project manager to get a feel for the project because a clear appreciation of the scale and complexity of the scheme helps to trigger the natural human instincts of when things look right or look wrong. These instincts are not simply based on a hunch but also on experience and familiarity with the construction process.

It is just as important to appreciate the financial scale of the project as well. Experience will often identify the project value or rate of expenditure as being inconsistent with the time allowed. To get a feel for the project, consider the following:

- Study the drawings and project documentation
- Visit the site

- Assess the scale and scope of the project
- Assess the approximate value of the project
- Consider the rate of spend (i.e. the relationship between value and time)

Establishing key project dates

The overall parameters of a project will be determined by establishing the key dates. Some of these will be established by the client and his advisors and conveyed to the contractor in the tender documentation, and others will be common sense. They will include:

- Project start and finish dates
- Sectional or phased completion dates
- Holiday periods
- Commissioning or handover

Establishing key activities or events

The next step is to determine the key activities or tasks to be carried out, together with any important events which should be included in the programme.

Activities are tasks or jobs to be done which have a time value. Obtaining planning permission is an activity because it can take several months, or even years, to complete. Events are points in time by which things must happen and have no time value. For example, the start of construction work is an event which triggers a series of activities which do have a time value.

The activities on the programme will vary according to the stage of the project. Some suggestions are listed here.

Check-list of key activities/events during the design stage

- Brief the design team
- Make professional appointments
- Notify the Health and Safety Executive (Form F10)
- Start the pre-construction health and safety plan and the health and safety file
- Obtain planning permission
- Apply for building regulation approval
- Prequalify contractors
- Appoint contractor(s)

Check-list of key activities/events during the tender stage

- Prepare tender documents
- Complete the pre-tender health and safety plan
- Period for tendering
- Evaluate tenders
- Compile contract documentation

Check-list of key activities/events during the pre-contract stage

- Appoint project staff
- Develop the construction health and safety plan
- Prepare requirement schedules
- Pre-start meeting
- Check the construction stage health and safety plan
- Permit start of construction work (the client has a statutory duty under CDM regulation 10)

Check-list of key activities/events during the construction stage

- Set up site establishment
- Groundworks or substructure
- Frame/external envelope
- Floors
- Roof structure and cladding
- Building watertight
- Mechanical, electrical and ventilation installations
- Finishes
- External works and drainage
- Practical or substantial completion
- Clear site

Assessing how long the activities will take

Assessing the duration of activities is not an exact science. At the early stages of a project, when little detailed information is available, a great deal of reliance must be placed on judgement and experience. Later on, calculations can be performed by considering the relationship between the quantity of work to be done and the output or rate of production anticipated. For example:

$$\frac{\text{Quantity}}{\text{Output per hour}} = \textbf{Hours}$$

$$\frac{\text{Hours}}{\text{No of hours per day (8)}} = \textbf{Days}$$

$$\frac{\text{Days}}{\text{No of days per week (5)}} = \textbf{Weeks}$$

Check-list

- The RIBA Plan of Work gives detailed information on the tasks undertaken by members of the design team from which time assessments can be made
- Many local authorities provide information and statistics relating to planning applications and the time taken to deal with them
- The Joint Consultative Committee for the Building Industry publishes codes of procedure which give indicative time allowances for tendering under various procurement arrangements
- EC tendering rules (see Chapter 1) establish times for the notification, tender periods and evaluation of public sector tenders
- The RICS Building Cost Information Service (BCIS) provides detailed tender analyses and contract periods for a wide variety of construction projects
- Any number of estimating books provide average outputs for a wide variety of building and civil engineering work. The quantity of work divided by the output per day will give the required duration in days
- Contractors are the best source of information but this unfortunately is rarely published
- Inevitably, a degree of common sense and experience is invaluable in assessing activity durations and sometimes there is no substitute for an educated guess

Establishing the sequence

This is best done initially on paper because that makes it easier to see the big picture. If project management software is being used from scratch, it is advisable to type the list of activities and then cut and paste them into the correct order *before* putting in any logical links. An awful tangle can result if you get this wrong!

Check-list

- Prepare a list of operations/activities. These must be significant activities which must have a duration and resource implication
- Assess durations in days or weeks

- Consider the order of work and overlap between related operations. The project logic is important and this can be established by taking each activity in turn and asking:
  - what must precede this activity?
  - what must follow this activity? (dependency)
  - what can happen at the same time (concurrency)?
- Try at this stage to present the project logic in the form of a bar chart or linked bar chart, or preferably in the form of a simple arrow diagram or as a series of precedence relationships. A 'doodle diagram' for the whole project can be useful and this is illustrated in Figure 4.10
- Consider subactivities as each main activity may have a sequence of its own. Many project management software packages have the facility of subnetworks or subheadings which can be expanded or 'rolled up' (compressed) at the click of a computer mouse. For example, an operation/activity such as 'construct pile caps' may have subactivities such as:
  - Excavate pile cap
  - Cut off and trim piles
  - Blind base of pile cap
  - Fix formwork to base
  - Fix steel reinforcement
  - Set bolts in base
  - Place concrete
  - Strip formwork
  - Backfill working space
- Consider 'start to start' or 'finish to start' relationships and whether there need to be time constraints between activities in order to create overlapping or concurrency
- The programme must be realistic and achievable

Deciding which programming technique to use

There is no strict rule as to which programming technique should be employed. This needs to be considered in the light of the size and complexity of the project in hand, any personal preferences and whether there are any stipulations in the contract documentation.

Bar charts are the easiest to use but they can give misleading results because there is no strict logic imposed on the programme. It may be better to use linked bar charts or arrow or precedence diagrams to overcome this problem. However, for repetitive work such as housing projects, line of balance may be preferred, or for roadworks, tunnelling or repetitive civil engineering work, time-chainage diagrams could be the best application to use.

When using project management software packages, it is usually advisable to draw out the programme on paper first. This helps to establish the correct logic and avoid the possibility of getting in a tangle when working at the computer screen – which is particularly likely to happen when making the logical links between activities as it is not always possible to see the whole picture on screen.

4.9 Bar charts and linked bar charts

History and development

Kempner (1980) and other management writers recognise that Henry Gantt first introduced bar charts for ship building projects in the early 1900s by popularising the graphical presentation of work versus time. Gantt belonged to the Scientific Management school of thinking of the late nineteenth and early twentieth centuries which included Taylor, Fayol and Gilbreth, among others. Gantt charts were the first scientific attempt to consider work scheduling against time. They have now become the basis of the modern bar chart, which has several variants.

Before the advent of powerful modern computers, bar charts were prepared by hand, often using pre-prepared blank sheets containing a column for the project activities and squares for drawing the bar lines. It was quite common to see the bar chart on the site agent's wall in the site cabin, with progress coloured in with crayon/coloured pencil and a vertical string pinned in place to denote the current date.

Limitations in the power and flexibility of bar charts resulted in the adoption of more sophisticated techniques in the 1950s and 1960s, in particular network analysis using arrow and later precedence diagrams. However, in more recent times, the development of the linked bar chart technique has led to a resurgence in the use of bar charts, mainly as a consequence of developments in project management computer software. There are a number of project management software packages available which employ linked bar charts as the preferred display. These include:

- C S Project Professional – Crest Software
- Hornet Windmill – Claremont Controls
- Microsoft Project – Microsoft Corporation
- Power Project Professional – Asta Development
- Project Commander – Construct-it USA

Principles of bar charts and linked bar charts

The bar chart is laid out with the time-scale in days/weeks/months/

years along the top axis and a list of tasks or activities down the left hand side. The time required for each activity is represented by a horizontal line (or bar), with the length of the line indicating the duration of the activity.

The software packages available for producing bar charts require the user to enter a logic by creating links between the activities on the programme, which overcomes the major disadvantage of Gantt charts. Users are able to present bar charts in a professional manner at any level of the planning process, and programmes can be readily updated, even on site, with the aid of lap-top computers.

One of the problems with the traditional Gantt charts is the tendency for the planner/manager to work backwards from the construction period stated in the contract documents in establishing the detailed programme. The resulting bar chart is often no more than wishful thinking. It has been said that the bar chart often suffers from a 'morning glory complex' – it blooms early in the project and is nowhere to be seen later on! This problem is overcome to some extent with the linked bar chart as this form of display forces the programmer to think about the logic of the programme as each link is developed.

Bar charts are well suited to depicting construction sequences and are readily understood at all levels of management. They can be used to develop the programme prepared at the tender stage into the master programme and likewise into the short-term planning throughout the contract period. Bar charts are easily and readily updated at weekly and monthly intervals. A colour coding system may be introduced for progress recording but most software packages have facilities for project tracking and updating which provides an accurate record of progress on the contract for future reference.

Most architects and site managers tend to have problems in understanding anything other than bar chart displays, even where network analysis has been stipulated in the contract documents as the project planning technique to be employed. However, bar charts have limitations and a key disadvantage is that they do not show dependency. Consequently, it is not easy to see the interrelationship between activities and how dependent they might be on one another. This is not so bad on a simple project but, where there is a large number of activities on the programme, real problems can arise for the manager. The problem can be overcome by using the linked bar chart.

Figure 4.4 illustrates the relationships used on linked bar charts. These include:

- Finish to start relationships
- Start to start relationships (overlaps)
- Finish to finish relationships

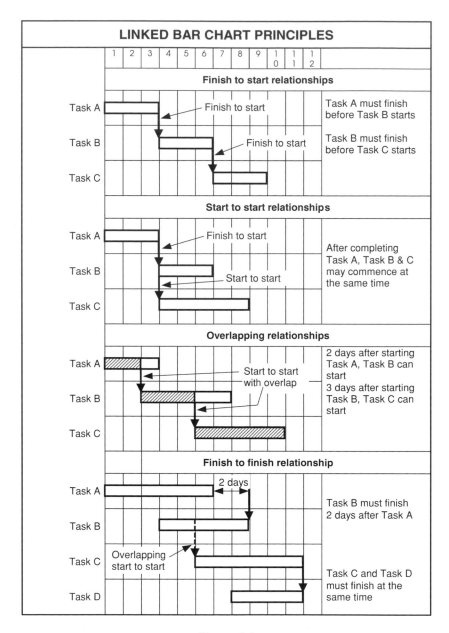

Figure 4.4

Advantages of bar charts and linked bar charts

- Simple format readily understood at all levels of management
- Applicable at all stages of the planning process: project planning, pre-tender, and pre-contract and contract planning

- Particularly useful in showing the relationship between the pre-tender programme, master programme and short-term programmes
- Clearly mimics the construction sequence – the use of linking between bars facilitates the overlapping of related operations
- Easily updated at weekly and monthly intervals for review purposes and progress reports
- Key milestone symbols may be introduced to highlight critical dates with regard to key contract stages, information requirements and as an aid to subcontractor and materials procurement
- Resources may be shown on the bar chart, which clearly relates labour, plant and subcontractors to the rate of working and helps the manager to see problems of continuity and waste
- Facilitates the production of labour histograms, value–time forecasts, cumulative labour and plant forecasts and other project budgets so that actual progress may be charted against planned progress
- Enables the contractor to quickly and simply fulfil contractual requirements to submit a programme prior to commencement of work on site
- Readily updated with information such as key site deliveries, progress to date and delays, making it simple to produce an 'as built' programme (or 'programme of the day') which may prove an asset to the contractor in forming contractual claims. The effect of the receipt of late information on programmed operations can be monitored and reported. Copies of the contract position at the date of a specific occurrence may provide evidence of the resulting delay
- The bar chart programme can be used to form the basis of financial forecasting for both the client and the contractor
- Computer printouts in full colour can be scaled down to A4 size, which is invaluable for management reports and ease of handling

Disadvantages of bar charts and linked bar charts

- Gantt charts do not show dependency and therefore do not clearly indicate which operations directly relate to the successful completion of the project. This makes it difficult to apply management by exception
- Logical links are used to overcome this problem but these can become confusing and difficult to interpret on complex projects
- Consequently, complex interrelationships cannot be clearly shown

Developing a linked bar chart

Creating a useful and professional-looking bar chart using modern software packages is a relatively straightforward task requiring a couple of hours to learn the package and then it is just a question of adopting a logical and systematic approach. Of course, it does help to know how buildings are put together as well!

One of the common packages which is simple to use, relatively inexpensive and gives good results is Microsoft Project. Figure 4.5 illustrates the following three basic steps in developing a simple programme using this package.

- *Step 1*
  Decide on the most appropriate time-scale for the programme. Months on the major scale and weeks on the minor scale are usually best for most programmes. For short-term planning and short duration projects, where a finer level of detail is necessary, weeks and days would be more appropriate. Next, list the various tasks/activities that make up the project and sort these into order using the cut and paste facility. Choose the activities carefully. Too few activities will make the programme of little use, whereas too many will be cumbersome and difficult to print out and read. Think about the possibility of using rolled-up tasks (called summary tasks in MS Project) which can be expanded into subtasks at a later stage. Note that the computer defaults to a duration of 1 day for each activity.

- *Step 2*
  Insert summary tasks and add durations (weeks are generally best) to the activities by calculation or from experience. Make any changes to the number or order of the activities before going on to step 3.

- *Step 3*
  Add logic to the programme by linking relevant activities to one another. Think about which activities must come first, which must follow and which may happen at the same time. Overlaps or delayed starts can be introduced by choosing the appropriate relationship from the menu (e.g. start to start) and by adding the necessary lag time.

Further features are easily added such as holiday periods, key milestones (events), resources and cost information.

Practical applications

The following examples illustrate the wide use of bar charts and linked bar charts for producing programmes for both client and contractor planning purposes. The examples shown illustrate the use and versatility of the bar chart/linked bar chart for providing a clear visual overview of the project at various stages.

DEVELOPING A BAR CHART (Using Microsoft Project)

Step 1

ID	Task Name	Duration
1	SET UP SITE	1d
2	BASEMENT PILING	1d
3	RC SLAB	1d
4	RC FRAME	1d
5	GROUND SLAB	1d
6	RC COLUMNS & BEAMS	1d
7	FLOORS	1d
8	PCC CLADDING	1d
9	WINDOWS	1d
10	ROOFING	1d
11	M&E	1d
12	FINISHES	1d
13	DRAINAGE	1d
14	ROADS & PAVINGS	1d
15	LANDSCAPING	1d
16	CLEAR SITE	1d

Step 2

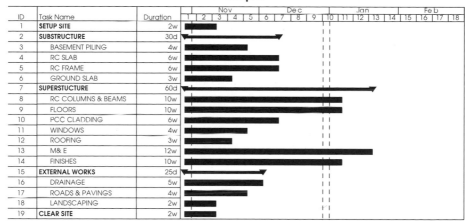

ID	Task Name	Duration
1	SETUP SITE	2w
2	SUBSTRUCTURE	30d
3	BASEMENT PILING	4w
4	RC SLAB	6w
5	RC FRAME	6w
6	GROUND SLAB	3w
7	SUPERSTUCTURE	60d
8	RC COLUMNS & BEAMS	10w
9	FLOORS	10w
10	PCC CLADDING	6w
11	WINDOWS	4w
12	ROOFING	3w
13	M& E	12w
14	FINISHES	10w
15	EXTERNAL WORKS	25d
16	DRAINAGE	5w
17	ROADS & PAVINGS	4w
18	LANDSCAPING	2w
19	CLEAR SITE	2w

Step 3

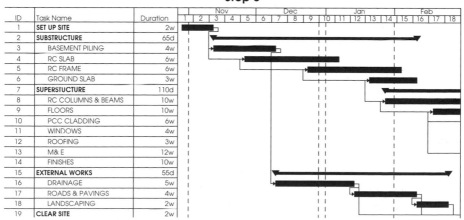

ID	Task Name	Duration
1	SET UP SITE	2w
2	SUBSTRUCTURE	65d
3	BASEMENT PILING	4w
4	RC SLAB	6w
5	RC FRAME	6w
6	GROUND SLAB	3w
7	SUPERSTUCTURE	110d
8	RC COLUMNS & BEAMS	10w
9	FLOORS	10w
10	PCC CLADDING	6w
11	WINDOWS	4w
12	ROOFING	3w
13	M& E	12w
14	FINISHES	10w
15	EXTERNAL WORKS	55d
16	DRAINAGE	5w
17	ROADS & PAVINGS	4w
18	LANDSCAPING	2w
19	CLEAR SITE	2w

Figure 4.5

Software packages are used extensively in practice; these allow colourful displays to be shown to aid presentation and include facilities for linking, adding milestone symbols and incorporating progress recording routines. Basic presentation techniques are aimed at simplifying the programme display to make it easy to read and record progress.

Figure 4.6 shows a client's project programme, indicating the key activities during both the design and tender stages as well as highlighting the anticipated period for the construction phase.

It is essential for the client team to have a programme in order to aid control. While little firm information will be available at the very early stages, the client's project manager or lead consultant should nevertheless draw up this programme based on a realistic estimate of how long the design, tendering and construction phases of the project should take. The CIOB Code of Practice for Project Management (2002) calls this the project master schedule, presumably to avoid confusion with the contractor's master programme.

The project master schedule shows a number of features which should be noted:

- The time-scale in years and months rather than weeks
- The use of 'negative' time to indicate the pre-construction period
- The inclusion of holiday periods without which the overall duration would be overoptimistic
- The use of milestones to indicate key events in the programme (e.g. handover)
- The emphasis on the client team activities necessary to plan the design and tender stages, including obtaining statutory approvals to build
- The construction phase denoted by a single bar line
- The 'start on site' and 'handover' dates would be included in the contract appendix
- Intermittent activities shown with a dotted line

The 5-month construction period indicated on the client's master schedule in Figure 4.6 would be developed into a pre-tender programme by the contractor at the tender stage.

The contract start and finish dates or contract duration would be given in the contract appendix and the contractor would need to check that this was realistic before tendering. Figure 4.7 indicates a 5-month (22-week) pre-tender programme with the main construction activities and durations shown. Features of this programme include:

- The time-scale shown in months and weeks
- The use of a linked bar chart display to show logic and dependency

PROJECT MASTER SCHEDULE

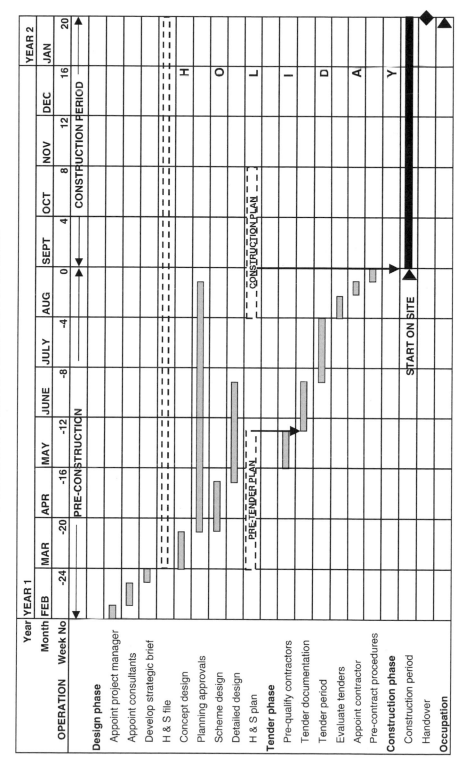

Figure 4.6

PRE-TENDER PROGRAMME

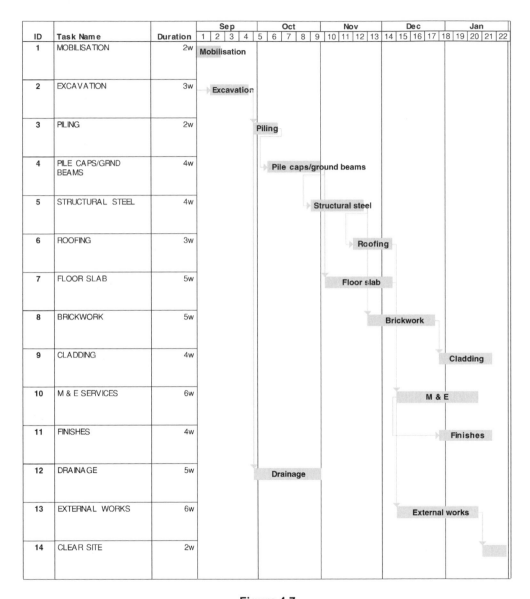

ID	Task Name	Duration
1	MOBILISATION	2w
2	EXCAVATION	3w
3	PILING	2w
4	PILE CAPS/GRND BEAMS	4w
5	STRUCTURAL STEEL	4w
6	ROOFING	3w
7	FLOOR SLAB	5w
8	BRICKWORK	5w
9	CLADDING	4w
10	M & E SERVICES	6w
11	FINISHES	4w
12	DRAINAGE	5w
13	EXTERNAL WORKS	6w
14	CLEAR SITE	2w

Figure 4.7

- The use of finish to start and start to start links between the tasks
- Activity/task names repeated on the bar line for ease of reading

At the pre-contract stage, the contractor would develop the pre-tender programme into his master programme, as shown in Figure 4.8.

MASTER PROGRAMME

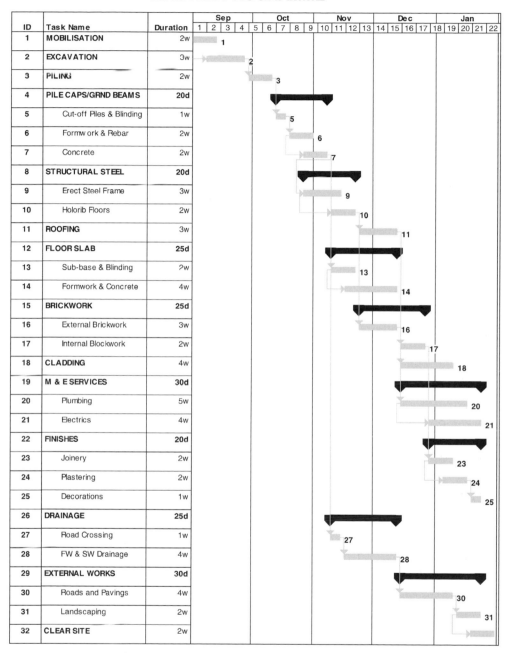

ID	Task Name	Duration
1	MOBILISATION	2w
2	EXCAVATION	3w
3	PILING	2w
4	PILE CAPS/GRND BEAMS	20d
5	Cut-off Piles & Blinding	1w
6	Formwork & Rebar	2w
7	Concrete	2w
8	STRUCTURAL STEEL	20d
9	Erect Steel Frame	3w
10	Holorib Floors	2w
11	ROOFING	3w
12	FLOOR SLAB	25d
13	Sub-base & Blinding	2w
14	Formwork & Concrete	4w
15	BRICKWORK	25d
16	External Brickwork	3w
17	Internal Blockwork	2w
18	CLADDING	4w
19	M & E SERVICES	30d
20	Plumbing	5w
21	Electrics	4w
22	FINISHES	20d
23	Joinery	2w
24	Plastering	2w
25	Decorations	1w
26	DRAINAGE	25d
27	Road Crossing	1w
28	FW & SW Drainage	4w
29	EXTERNAL WORKS	30d
30	Roads and Pavings	4w
31	Landscaping	2w
32	CLEAR SITE	2w

Figure 4.8

4.10 Network analysis – The Critical Path Method

History and development

In 1956, the E. I. du Pont de Nemours Company established a team to study new management techniques for the company's engineering functions. One of the first areas to be considered was the planning and scheduling of construction work. Data was input into a UNIVAC 1 computer in the form of construction sequences and activity durations in order to generate a schedule of work.

In early 1957, J. W. Mauchley, J. E. Kelley Jr and M. Walker developed the basic principles of the Critical Path Method (CPM). A test group was set up to apply the new technique to a chemical plant project in Kentucky. In 1958, Mauchley Associates developed a series of training programmes to spread knowledge and use of the method throughout industry. The development of PERT (Program Evaluation and Review Technique) was originated by the Special Projects Office for the Polaris missile programme.

In the UK, development work on CPM was undertaken by the Building Research Establishment and a number of papers were published by Nuttall and Jeans (1960a,b).

The initial applications developed were only suitable for large, complex projects and required the assistance of main frame computers to analyse the data. Data input and output data retrieval were slow and difficult to interpret – in fact the progress of the project had often changed by the time the data was analysed and reported on. Reece (1989), in his article on the future of project planning, reviewed the rise and fall of CPM. The resurgence in the use of CPM in the late 1980s was mainly due to the introduction of the personal computer and developments in user-friendly project planning software. Now, linked bar charts are more popular for most projects but, for the larger schemes, CPM still has an important part to play, especially with the advent of cheaper software and hardware.

There is little doubt that networks are here to stay, albeit that developments in project management software using precedence and linked bar chart presentation formats are largely preferred to arrow diagram techniques.

Whatever the case, as construction projects become larger and more complex to plan and manage, project managers and the larger contractors will become increasingly reliant on the computer to undertake project planning tasks, and the use of project management software will become the norm. Of course, the computer has now become an everyday site tool for both the construction manager and planner.

Project management-based software is regularly reviewed in the CIOB *Construction Computing Journal* and some of the software available in CPM format at the time of writing includes:

- Micro Planner Expert – Micro Planning International
- Plantrac – Computerline Ltd
- Primavera – Forgetrack Ltd
- Superproject – Computer Associates

Principles of networks

The Critical Path Method is presented in the form of an arrow diagram. The distinguishing feature of these diagrams is that the arrow represents the activity and the circle or node between the arrows is the event. By numbering the events, the arrow activities can be identified. This is usually done in numerical order starting at the beginning of the network and progressing to the end, ensuring that the number at the tail of the arrow is smaller than that at the head.

Each activity is given a duration, and the earliest and latest event times of the activity can be calculated by making forward and backward passes through the network. These times are recorded in the node or event circles. From this information, a schedule can be produced which will facilitate calculation of the total float or spare time for each activity. Dummy activities, which usually have no duration or value, can be introduced to indicate dependencies not shown by the arrow activities.

The project management software carries out the time analysis for the programme sequence developed by the planner. This allows the operations to be sorted into priority, thereby enabling the critical path to be calculated and highlighted. The critical path is defined as the longest route through the sequence of operations which must be undertaken in order to complete the project.

Advantages of networks

- Through the discipline of CPM, the user can achieve better planning due to the logical approach undertaken during the development of the construction sequence
- Identification of critical operations on which effort and resources can be applied aids the contractor's management. This enables 'management by exception' to be applied to critical activities
- CPM allows the planner to express his ideas in graphical form. Recent developments in project management software allow networking on the screen to be carried out
- The planner has the facility to assign priorities for labour, plant, material and subcontractor resources to each operation on the network
- Bar chart analysis aids understanding at site management level for both the contractor and client

- The effect of changes and variations can be evaluated and time–cost optimisation analysis undertaken
- Cash flow assessments and valuation forecasting information may be output in graphical format and readily updated during the project
- Using laptop and site-based computers, progress may be speedily analysed and a variety of management reports made available

Disadvantages of networks

- The development of a network sequence using arrows joined at node points is cumbersome and does not easily facilitate concurrent activities being shown
- Activities can be split into stages or sections to overcome this problem but this does not convincingly represent the situation on the project
- Often in practice several activities are scheduled to start just after the preceding one, so there is a complex concurrency but with a delay at the beginning. This can be overcome by using a ladder diagram. This introduces concurrent activities with a 'lead–lag' start and finish which can be likened to dummy activities but with a duration attached
- In these situations precedence diagrams offer much more flexibility
- For all the man-hours spent on in-company management training programmes learning to appreciate networks, they have never captivated managers in small and medium-sized companies

Developing an arrow diagram

Micro Planner software uses arrow diagrams (it can also switch to precedence and bar chart format) and the 'Professional' version is a powerful project management tool. The software uses a 'drag and drop' technique which is relatively easy to use. This technique is particularly valuable for beginners to CPM and the arrow network helps to impose the discipline of strict logical thinking before graduating to other methods.

The following example illustrates the process of developing an arrow diagram.

Example – site layout

Figure 4.9 shows the plan of a site compound layout which is to be set up at the commencement of a project. A schedule of activities has been abstracted from the drawing and listed in the table. The activities have

SITE LAYOUT PLAN – ESTABLISHING THE SITE

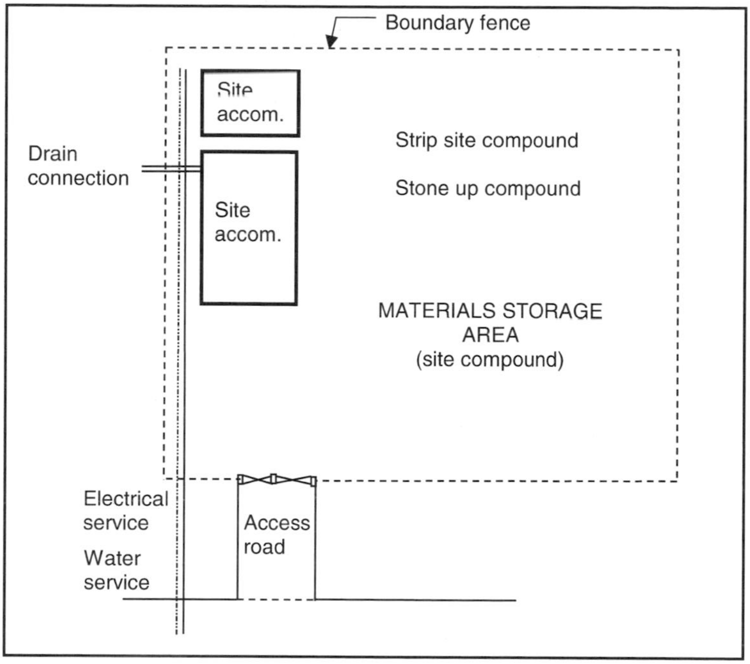

Boundary fence

Site accom.

Drain connection

Strip site compound

Stone up compound

Site accom.

MATERIALS STORAGE AREA
(site compound)

Electrical service

Access road

Water service

Operation	Duration (days)
Excavate and stone up access road	3
Erect fence	4
Strip site compound	4
Stone up the compound	2
Erect site accommodation	5
Fit out	5
Electrical service	1
Temporary drain connection	3
Excavate and lay water service	3
Fit gates	1

Figure 4.9

not been listed in any specific order. A practical assessment of the duration of each activity has been made in days.

The basic thought process in developing a network diagram for the project is as follows:

- *Stage 1* – Assess the overall construction period
  This involves a study of the project drawings in order to get a feel for the project. In order to assess the overall time period, ask the

question, 'Is the duration likely to be 3 weeks, 5 weeks or 100 weeks?' Normally this will be based on experience or by consulting somebody in the organisation who does know.

- *Stage 2* – Identify the construction operations
  From the project drawings a list or schedule of operations will be drawn up. This should cover in sufficient detail the building sequence appropriate to the stage of planning being carried out. For example, a pre-tender programme may only relate to the major operations or work stages, whereas a short-term programme will cover more detailed operations or sub-operations.

- *Stage 3* – Assess activity durations
  For each operation an assessment should be made of its appropriate duration (in days or weeks). Durations may be based on the experience of the manager/planner or on an analysis of the man-hour allocation in the estimate.

- *Stage 4* – Establish sequence of work
  An attempt should now be made to express the operations/activities in the form of an initial arrow diagram (or 'doodle' diagram) in order to commit pen to paper. Figure 4.10 indicates the initial network sequence. When developing an initial arrow diagram the following questions should be posed:
  o   Which is the first activity in the sequence?
  o   Which activities must be completed before the next one starts? (finish to start relationship)
  o   Which activities can be undertaken at the same time?
  o   Which is the last activity in the sequence?
  It is important at this stage that the arrow diagram is based on finish to start relationships and it is important to check the operational logic to ensure that the sequence of construction is correct. It is no good planning to put the roof on a building until all the columns supporting the roof have been erected. Figure 4.11 indicates a redrafted arrow diagram containing all the relevant data:
  o   Event numbers
  o   Earliest event times
  o   Latest event times
  o   Floats
  o   The identification of the critical path.

Below the arrow diagram, a time-scaled bar chart has been developed from the analysed data. The presentation of the bar chart in this form allows the critical operations to be presented in a single horizontal bar line, with the non-critical activities grouped together. This approach

ESTABLISHING SITE

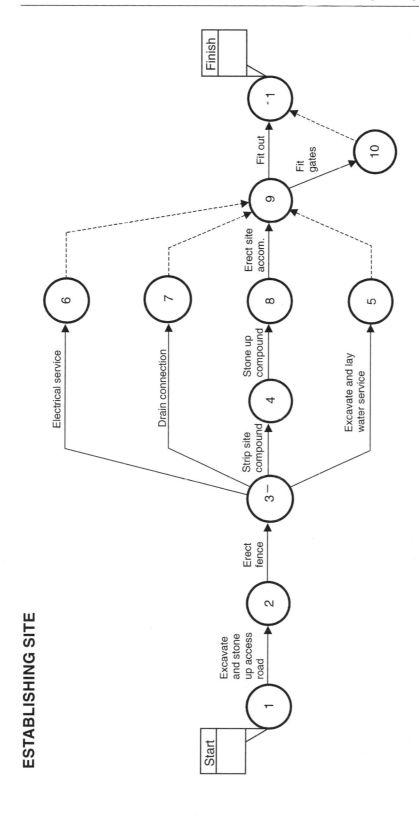

INITIAL NETWORK DIAGRAM
(Doodle diagram)

Procedure – Establish logic
Draft arrow diagram
Enter event numbers

Figure 4.10

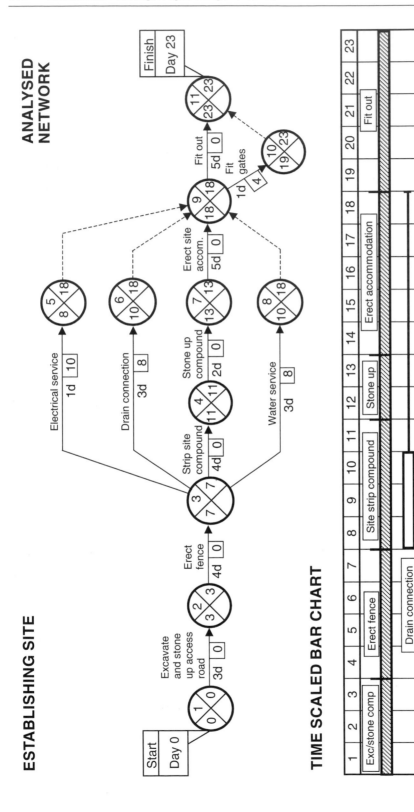

Figure 4.11

simplifies the presentation for the understanding of site management personnel.

Practical applications

Example – cumulative valuation forecast

The following example illustrates the use of arrow diagrams to prepare a cumulative valuation forecast for a sewage treatment works project at both the earliest and latest event times. The steps involved are as follows:

- *Step 1*
  Prepare an initial arrow diagram for the project together with a list of the monetary values of each of the activities (Figure 4.12).

- *Step 2*
  Analyse the arrow diagram in order to calculate the overall project duration of 22 weeks (Figure 4.13).

- *Step 3*
  Prepare separate bar charts showing the programme using both the earliest event times and latest event times for the project (Figures 4.14 and 4.15).

- *Step 4*
  Allocate monetary values to the activities on each bar chart programme. Calculate the cumulative value for each situation and indicate the valuation periods (4-weekly in this example).

- *Step 5*
  Calculate the cumulative monthly values for each bar chart situation and draw the respective value–time line graphs based on the earliest and latest event times (Figures 4.16 and 4.17).

- *Step 6*
  Combine the graphs to give a value–time 'envelope' based on earliest and latest event times (Figure 4.18).

The cumulative valuation envelope can be used to monitor the actual value release on the contract, which may form part of the contractor's monthly budgetary control procedures (see Chapter 9, Figure 9.3).

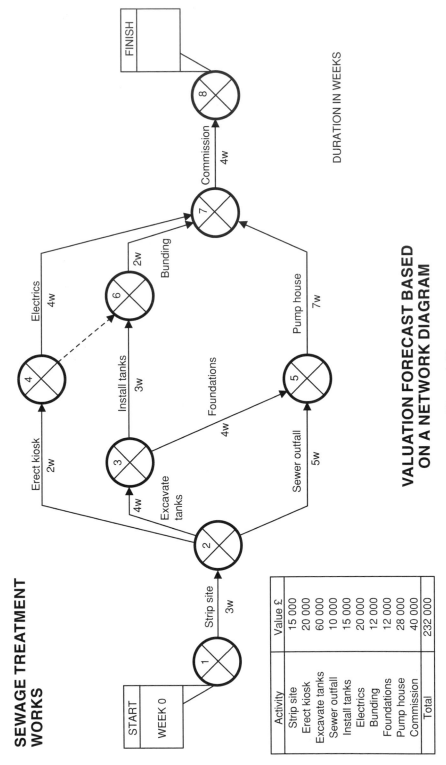

SEWAGE TREATMENT WORKS

Activity	Value £
Strip site	15 000
Erect kiosk	20 000
Excavate tanks	60 000
Sewer outfall	10 000
Install tanks	15 000
Electrics	20 000
Bunding	12 000
Foundations	12 000
Pump house	28 000
Commission	40 000
Total	232 000

DURATION IN WEEKS

VALUATION FORECAST BASED ON A NETWORK DIAGRAM

Figure 4.12

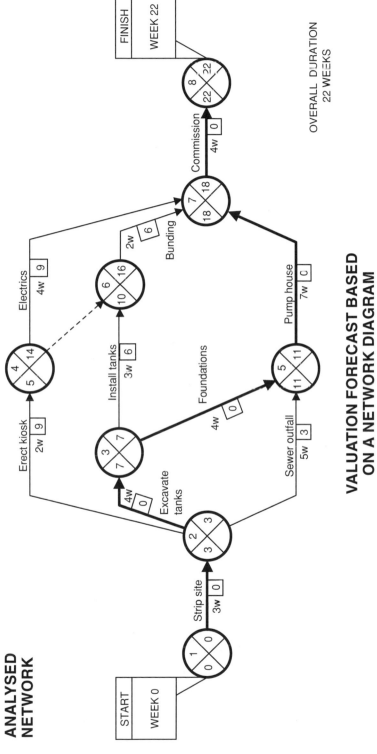

**VALUATION FORECAST BASED
ON A NETWORK DIAGRAM**

Figure 4.13

EARLIEST EVENT TIME – BAR CHART PROGRAMME

Time in weeks

OPERATION	ES	Dur	FL	1	2	3	4	5	6	7	8	9	10	11	12	13	14	15	16	17	18	19	20	21	22	VALUE
Strip site	0	3	0	5	5	5																				15000
Excavate tanks	3	4	0				15	15	15	15																60000
Foundations	7	4	0								3	3	3	3												12000
Pump house	11	7	0												4	4	4	4	4	4	4					28000
Commission	18	4	0																			10	10	10	10	40000
Erect kiosk	3	2	9				10	10																		20000
Electrics	5	4	9						5	5	5	5														20000
Sewer outfall	3	5	3				2	2	2	2	2															10000
Install tanks	7	3	6								5	5	5													15000
Bunding	10	2	6											6	6											12000
Weekly value				5	5	15	27	27	22	22	15	13	8	9	10	4	4	4	4	4	4	10	10	10	10	232000
Cumulative value				5	10	15	42	69	91	113	128	141	149	158	168	172	176	180	184	188	192	202	212	222	232	
Monthly value							42				128				168				184				212			232	
Valuation dates							1			2				3				4				5			6	

Figure 4.14

LATEST EVENT TIME – BAR CHART PROGRAMME

Time in weeks

OPERATION	LS	Dur	FL	1	2	3	4	5	6	7	8	9	10	11	12	13	14	15	16	17	18	19	20	21	22	VALUE
Strip site	0	3	0	5	5	5																				15000
Excavate tanks	3	4	0				15	15	15	15																60000
Foundations	7	4	0								3	3	3	3												12000
Pump house	11	7	0												4	4	4	4	4	4	4					28000
Commission	18	4	0																			10	10	10	10	40000
Erect kiosk	12	2	9													10	10									20000
Electrics	14	4	9															5	5	5	5					20000
Sewer outfall	6	5	3							2	2	2	2	2												10000
Install tanks	13	3	6														5	5	5							15000
Bunding	16	2	6																	6	6					12000
Weekly value				5	5	5	15	15	15	17	5	5	5	5	4	14	19	14	14	15	15	10	10	10	10	232000
Cumulative value				5	10	15	30	45	60	77	82	87	92	97	101	115	134	148	162	177	192	202	212	222	232	
Monthly value							30				82				101				162				212		232	
Valuation dates							1				2				3				4				5		6	

Figure 4.15

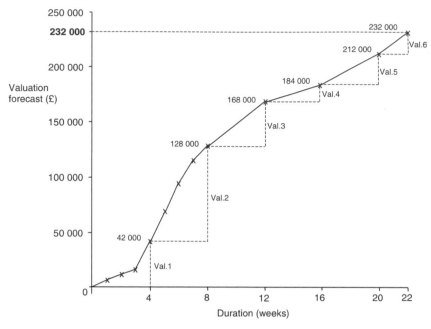

EARLIEST EVENT TIME VALUE FORECAST	
Week	Cumulative value
4	42 000
8	128 000
12	168 000
16	184 000
20	212 000
22	232 000

Figure 4.16

Example – labour resource allocation

The following example illustrates the principles of labour resource allocation applied to an arrow diagram sequence. It relates to the labour resourcing of the bricklaying operations in order to achieve continuity of work. The example shows the use of float times on non-critical activities in order to gain continuity of work for an individual trade.

- *Step 1*
  Prepare an initial arrow diagram for the project indicating the bricklaying operations H, C and M (Figure 4.19).

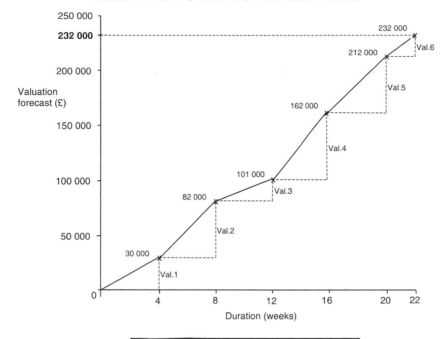

VALUE – TIME FORECAST
RELATIVE TO LATEST EVENT TIME

LATEST EVENT TIME VALUE FORECAST	
Week	Cumulative value
4	30 000
8	82 000
12	101 000
16	162 000
20	212 000
22	232 000

Figure 4.17

- *Step 2*
  Analyse the network indicating an overall project period of 25 days (Figure 4.20).

- *Step 3*
  Prepare the earliest event time bar chart highlighting the bricklaying operations. Allocate resources to the bar chart and draw the earliest start labour histogram showing the critical labour requirements

VALUE – TIME ENVELOPE
RELATIVE TO EARLIEST AND LATEST EVENT TIMES

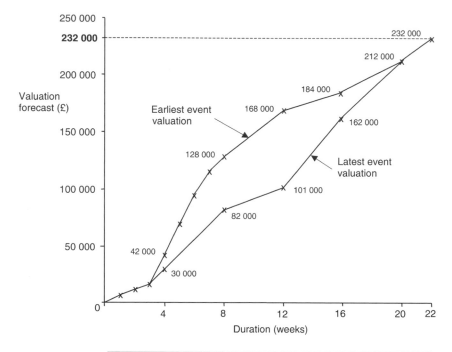

	EARLIEST EVENT TIME VALUE FORECAST	LATEST EVENT TIME VALUE FORECAST
Week	Cumulative value	Cumulative Value
4	42 000	30 000
8	128 000	82 000
12	168 000	101 000
16	184 000	162 000
20	212 000	212 000
22	232 000	232 000

Figure 4.18

shaded. In order to obtain continuity of work for the bricklayers it will be necessary to consider the available float on the non-critical brickwork operations H and M (Figure 4.21).

- *Step 4*
  Revise the bar chart programme using available float on operations H and M and level the resourced labour histogram (Figure 4.22).

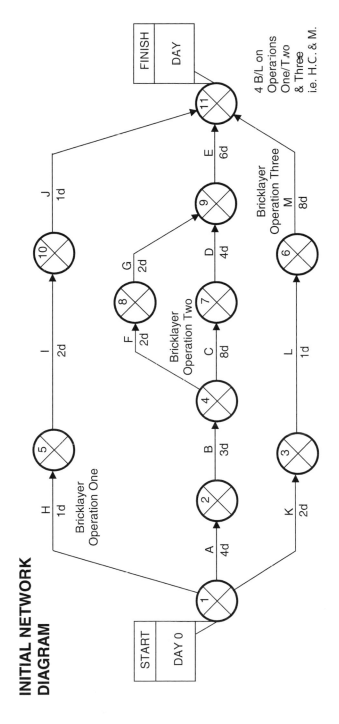

INITIAL NETWORK DIAGRAM

Bricklayer Operation One

Bricklayer Operation Two

Bricklayer Operation Three

START   DAY 0

FINISH   DAY

4 B/L on Operations One/Two & Three i.e. H.C. & M.

RESOURCING BRICKLAYING OPERATIONS TO ACHIEVE CONTINUITY OF WORK

Figure 4.19

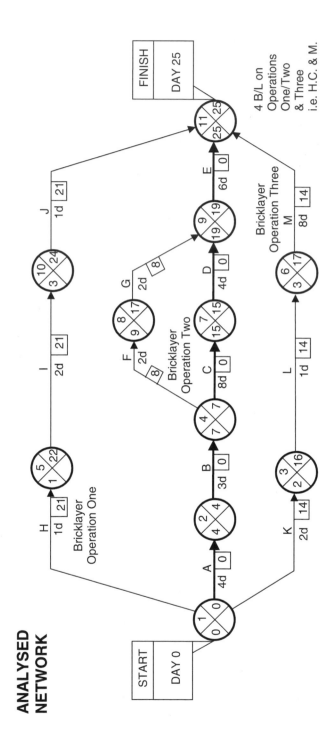

ANALYSED NETWORK

RESOURCING BRICKLAYING OPERATIONS TO ACHIEVE CONTINUITY OF WORK

Figure 4.20

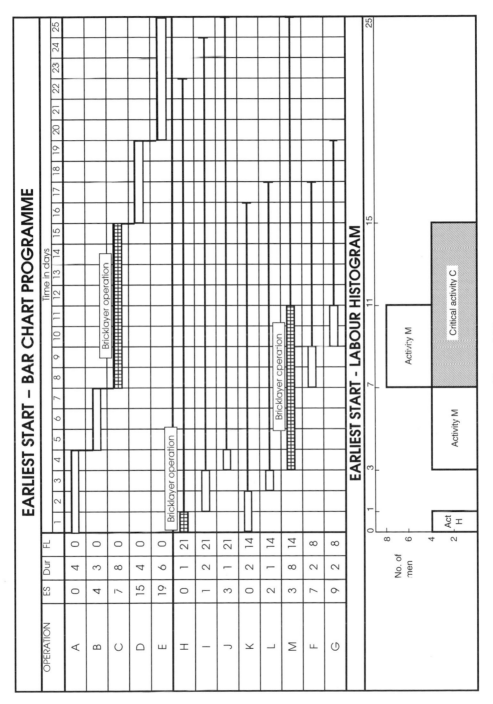

Figure 4.21

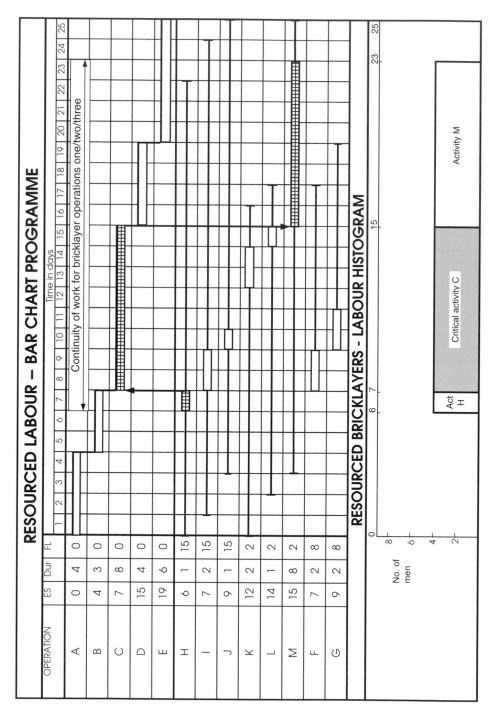

Figure 4.22

4.11 Precedence diagrams

History and development

Precedence diagrams follow the same logical procedures as arrow networks except that the activities and their dependencies are drawn differently. The precedence diagram consists of a series of boxes inter-linked with lines. The box or node represents the activity and the linking arrow indicates the relationships of the activities to one another. The box contains an activity label or name and duration. There is space for the earliest and latest start and finish times of the activity and a reference number may also be included if required.

Both the boxes and the lines may be given a time value. The time given in the box represents the duration of the activity, while any time on the line or arrow adds a dependency which might be a lead or a lag as required. Precedence diagrams do not require dummies to preserve the logic of the relationships and each node is ascribed a unique activity number.

Precedence diagrams were developed in the early 1970s by the Cementation Company as an alternative approach to network analysis which could more readily be applied to works of a civil engineering nature. In practice precedence diagrams are far more widely used than arrow networks because they are more flexible and more easily reflect the ways things happen.

One of the key reasons for the growth of precedence diagrams is the limitation of arrow diagrams when, for instance, one activity is required to start before the preceding activity is completed. This either means dividing the preceding activity into smaller parts or introducing a dummy with a time value.

The precedence approach introduced the idea of activity boxes, rather than activity arrows, which permits a number of different relationships to be expressed between activities. This approach relates more closely to the real situation on a construction project and this practicality makes the technique more popular. The relationships which can be included are:

- Finish to start
- Start to start
- Finish to finish
- Start to finish

This makes the precedence display easier to follow and permits the introduction of time constraints on the logical links without the need to include dummies or ladders.

PRINCIPLES OF PRECEDENCE DIAGRAMS

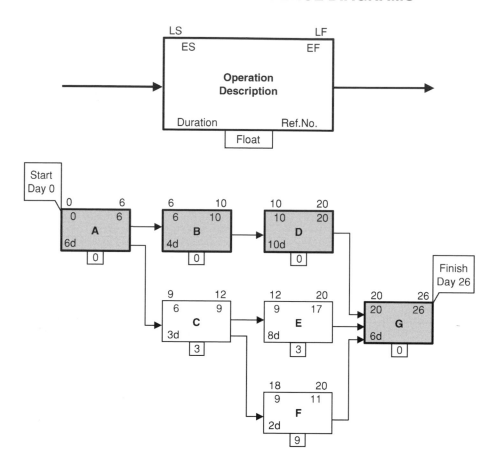

PRECEDENCE NOTATION

Figure 4.23

Principles of precedence diagrams

Figure 4.23 depicts the notation used in the precedence format, together with an example illustrating the basic principles of analysing a precedence sequence with the forward and backward calculations required to establish the critical path. Figures 4.24 to 4.28 illustrate the various relationships used when developing a precedence diagram. These have been shown in network (arrow), precedence, and linked bar chart formats in order to develop familiarity with the presentation.

It is important to understand clearly the relationships between arrow diagrams, precedence diagrams and bar charts so that they can readily be related to the construction process at site level. Using precedence

PRINCIPLES OF PRECEDENCE DIAGRAMS
(applied to factory project)

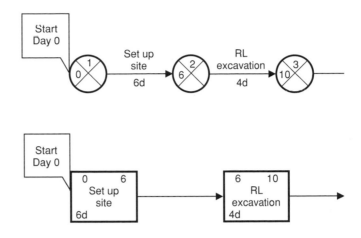

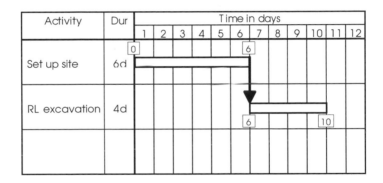

Activity	Dur	Time in days
		1 2 3 4 5 6 7 8 9 10 11 12
Set up site	6d	0 ... 6
RL excavation	4d	6 ... 10

FINISH TO START RELATIONSHIPS

Figure 4.24

diagrams, dummies are eliminated and the resulting 'number crunch-ing' analysis becomes relatively more simplified. The total float for each activity can be readily calculated from the diagram by deducting the earliest start time from the latest finish time in the corners of the activity box less the activity duration.

Figure 4.29 illustrates a precedence sequence for the early stages of a factory project. This clearly shows the overlapping of operations such as establishing the site, clearance, piling and the pile cap construction. The example has been used to illustrate start to start and lag-start relationships.

PRINCIPLES OF PRECEDENCE DIAGRAMS
(applied to factory project)

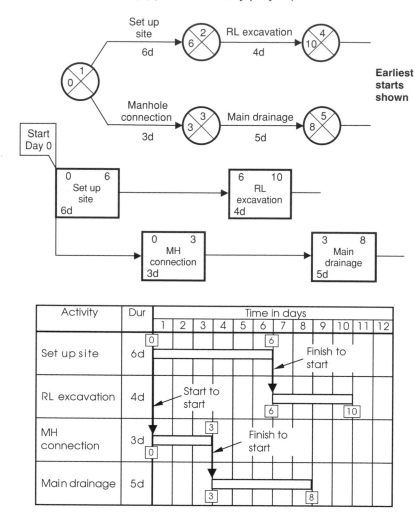

START TO START RELATIONSHIPS

Figure 4.25

The earliest and latest start and finish times are calculated by applying the 'forward and backward pass' routine to the precedence diagram. Where the times are the same this indicates the route of the critical path. Note that the start of a precedence activity may be on the critical path even though the activity itself may have float. This is illustrated graphically in the bar chart display underneath.

PRINCIPLES OF PRECEDENCE DIAGRAMS
(applied to factory project)

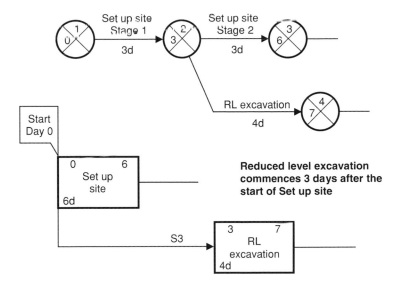

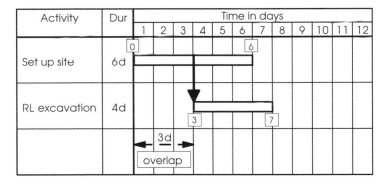

Reduced level excavation commences 3 days after the start of Set up site

START TO START RELATIONSHIPS
(Overlapping Activities)

Figure 4.26

Developing a precedence diagram

The process of developing a precedence diagram is illustrated in Figure 4.30 which shows a precedence diagram for establishing the site compound. All the relationships are finish–start. Start and finish flags have been entered on the diagram and the critical operations have been

PRINCIPLES OF PRECEDENCE DIAGRAMS
(applied to factory project)

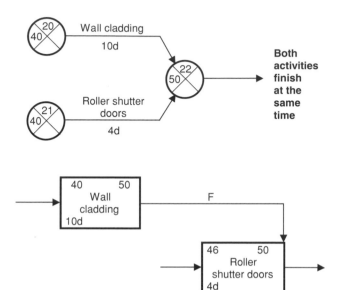

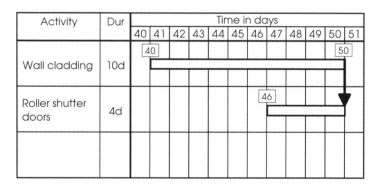

Activity	Dur	Time in days											
		40	41	42	43	44	45	46	47	48	49	50	51
Wall cladding	10d	40										50	
Roller shutter doors	4d							46					

FINISH TO FINISH RELATIONSHIPS

Figure 4.27

highlighted. The sequence has been presented in bar chart format in Figure 4.31. Compare this to the arrow diagram approach in Figure 4.11.

The use of precedence diagrams as a means of expressing construction relationships has now largely replaced arrow diagrams. This is mainly due to the simplicity of linking operations to each other in various ways and the ability to introduce time restraints realistically, conveniently and without overcomplicating the diagram.

PRINCIPLES OF PRECEDENCE DIAGRAMS

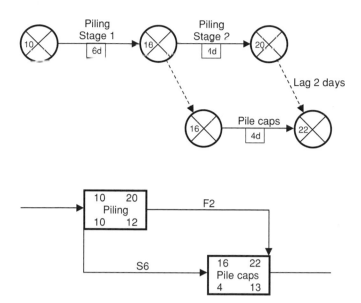

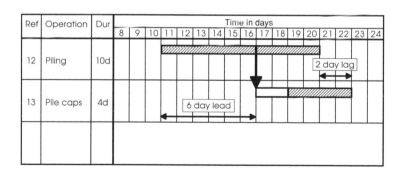

START TO FINISH RELATIONSHIPS

Figure 4.28

Project management software packages, such as Micro Planner Expert, enable a variety of links to be drawn between precedence boxes directly on the computer screen by using the 'drag and drop' technique. Software is also available which allows the planner/project manager to switch between precedence diagrams and linked bar chart displays in order to see how the overall programme is developing. Developments in computer software have had a significant impact on the development and understanding of the precedence diagram as a management tool.

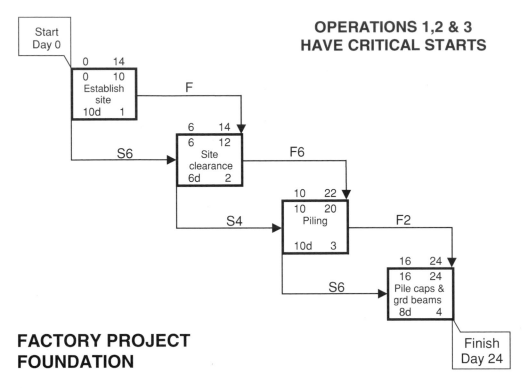

FACTORY PROJECT FOUNDATION

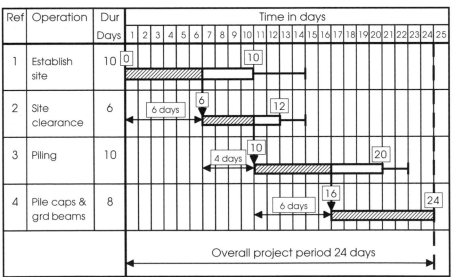

BAR CHART DISPLAY

Figure 4.29

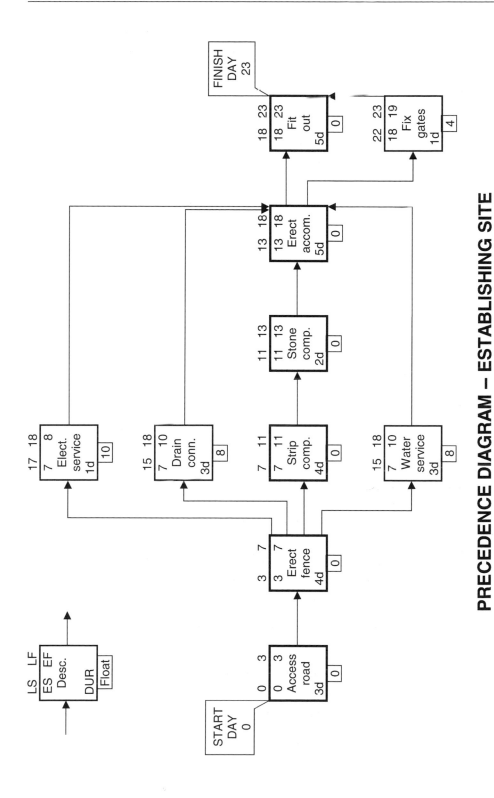

PRECEDENCE DIAGRAM – ESTABLISHING SITE

Figure 4.30

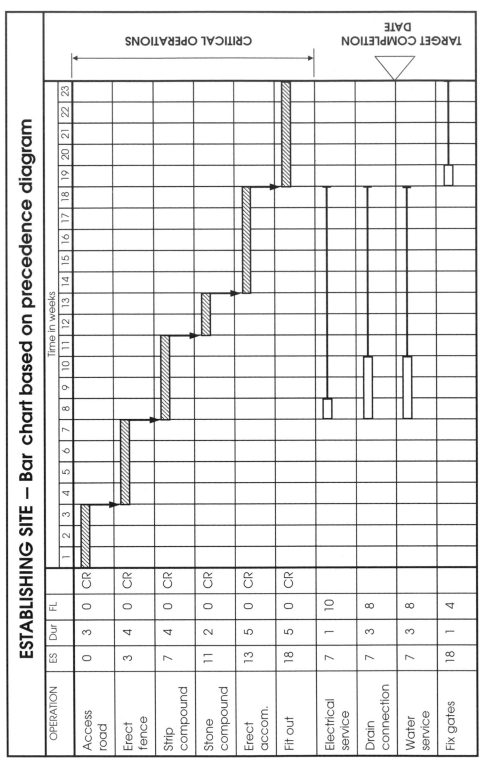

Figure 4.31

Practical applications

Example – precedence diagram

The following example illustrates the analysis of a precedence diagram based on a variety of start to start, finish to start and finish to finish relationships.

The sequence relates to the foundations, frame and ground floor slab operations during the construction of a factory building.

- *Step 1*
  Draw the initial network precedence diagram (Figure 4.32).

- *Step 2*
  Analyse the precedence diagram indicating floats and critical operations in order to achieve an overall completion date of 34 days (Figure 4.33). The objective of the exercise is to produce a cumulative project value forecast based on the earliest start situation.

- *Step 3*
  Draw an earliest start bar chart showing the critical operations listed first in the sequence. The analysis indicates that the start to start periods (overlaps) on the first three site operations are critical to achieving the overall completion date of 34 days. This has been highlighted on the bar chart by linking the critical operations (Figure 4.34).

- *Step 4*
  Allocate monetary values to the bar lines on the bar chart in order to produce the cumulative weekly value forecast (Figure 4.35).

- *Step 5*
  Draw the value–time forecast presented both graphically and in tabular format (Figure 4.36).

The cumulative value forecast, based on a realistic and achievable programme, enables the planning process to be integrated with the project monthly cost–value reporting procedures. Examples of analysing cost and value variances are covered in Chapter 9.

Example – precedence diagram (labour resources)

The following example illustrates the analysis of a precedence diagram based on finish–start relationships with the objective of producing a forecast of the labour resources at the earliest start situation, presented in both bar chart and histogram format.

PRECEDENCE DIAGRAM

INITIAL
PRECEDENCE
DIAGRAM

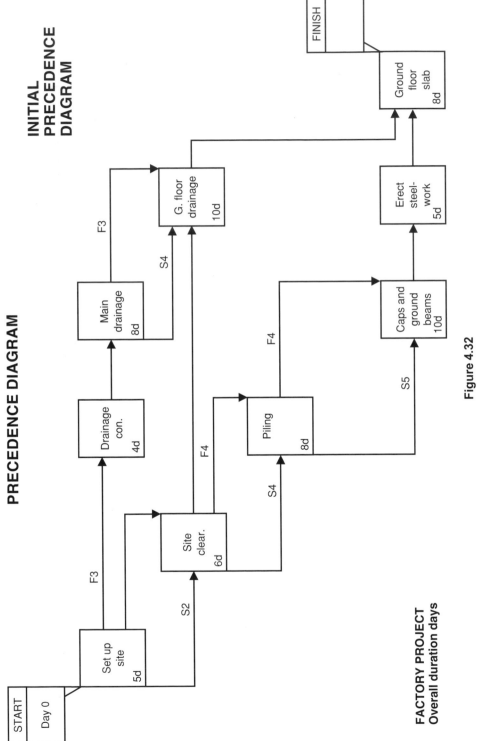

Figure 4.32

FACTORY PROJECT
Overall duration days

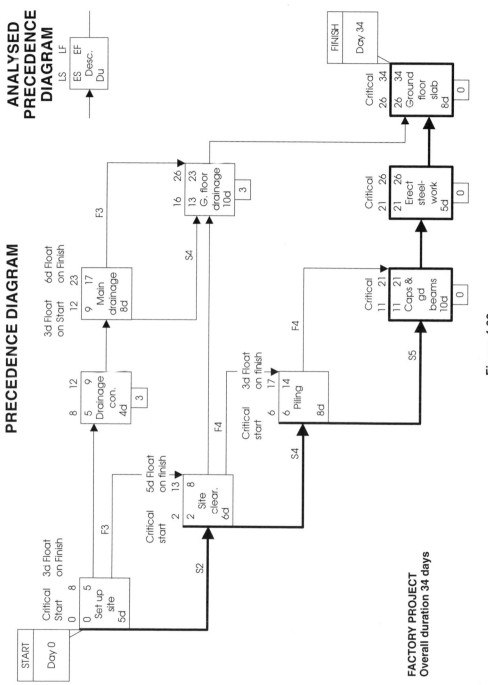

ANALYSED PRECEDENCE DIAGRAM

```
LS  LF
ES  EF
    Desc.
    Du
```

PRECEDENCE DIAGRAM

START
Day 0

Set up site
Critical Start
0 8
0 5
5d
3d Float on Finish

S2

F3

Drainage con.
8 12
5 9
4d
3

Main drainage
3d Float on Start   6d Float on Finish
12 23
9 17
8d

F3

Site clear.
Critical start   5d Float on finish
2 13
2 8
6d

F4

S4

G. floor drainage
16 26
13 23
10d
3

S4

Piling
Critical start   3d Float on finish
6 17
6 14
8d

F4

S5

Caps & gd beams
Critical
11 21
11 21
10d
0

S4

Erect steel-work
Critical
21 26
21 26
5d
0

Ground floor slab
Critical
26 34
26 34
8d
0

FINISH
Day 34

FACTORY PROJECT
Overall duration 34 days

Figure 4.33

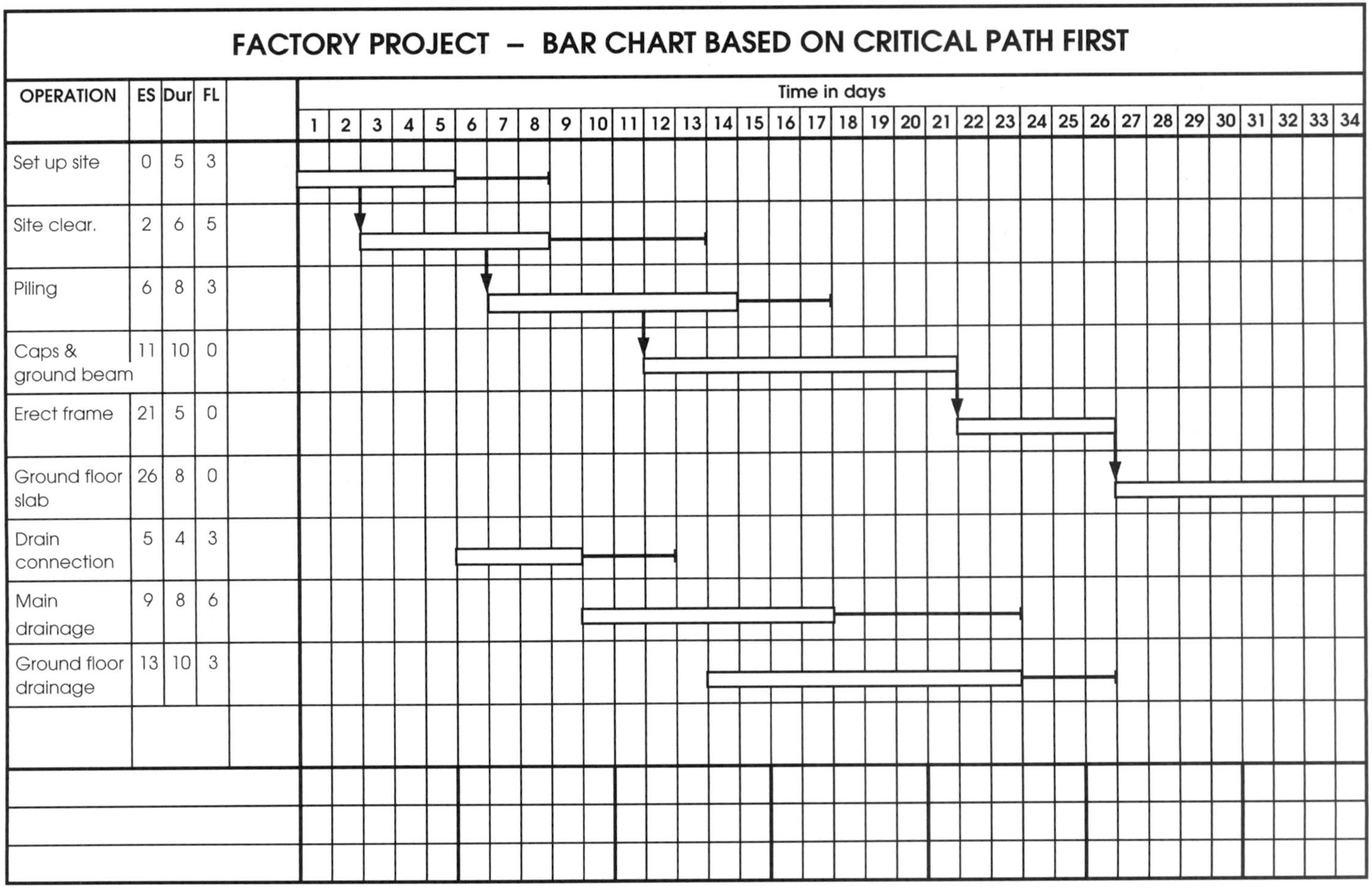

FACTORY PROJECT – BAR CHART BASED ON CRITICAL PATH FIRST

Time in days

OPERATION	ES	Dur	FL	1	2	3	4	5	6	7	8	9	10	11	12	13	14	15	16	17	18	19	20	21	22	23	24	25	26	27	28	29	30	31	32	33	34
Set up site	0	5	3																																		
Site clear.	2	6	5																																		
Piling	6	8	3																																		
Caps & ground beam	11	10	0																																		
Erect frame	21	5	0																																		
Ground floor slab	26	8	0																																		
Drain connection	5	4	3																																		
Main drainage	9	8	6																																		
Ground floor drainage	13	10	3																																		

Figure 4.34

FACTORY PROJECT – BAR CHART (from precedence diagram)

Time in days

OPERATION	ES	DV	FL	VALUE	1	2	3	4	5	6	7	8	9	10	11	12	13	14	15	16	17	18	19	20	21	22	23	24	25	26	27	28	29	30	31	32	33	34
Set up site	0	5	3	5000	1	1	1	1	1																													
Site clear.	2	6	8	12000			2	2	2	2	2	2																										
Piling	6	8	7	24000							3	3	3	3	3	3	3	3																				
Caps & ground beam	11	10	0	20000											2	2	2	2	2	2	2	2	2	2														
Erect frame	21	5	0	50000																						10	10	10	10	10								
Ground floor slab	28	8	0	16000																											2	2	2	2	2	2	2	2
Drain connection	5	4	3	4000						1	1	1	1																									
Main drainage	9	8	9	8000									1	1	1	1	1	1	1	1																		
Ground floor drainage	13	10	3	10000													1	1	1	1	1	1	1	1	1	1												
Prelims		£1000 per day		34000	1	1	1	1	1	1	1	1	1	1	1	1	1	1	1	1	1	1	1	1	1	1	1	1	1	1	1	1	1	1	1	1	1	1
Daily value					2	2	4	4	4	4	7	7	5	5	5	7	7	8	5	5	5	4	4	4	4	12	12	11	11	11	3	3	3	3	3	3	3	3
Cumulative value					2	4	8	12	16	20	27	34	39	44	49	56	63	71	76	81	86	90	94	98	102	114	126	137	148	159	162	165	168	171	174	177	180	183
Weekly value									16					28					32					22					50					23				12

Figure 4.35

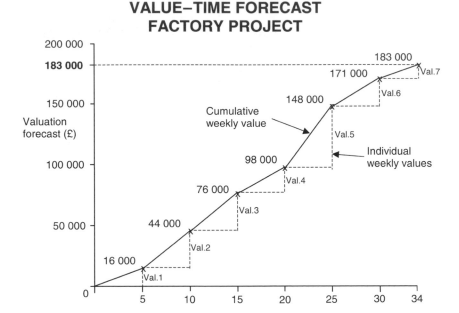

VALUE–TIME FORECAST FACTORY PROJECT

TABULAR CUMULATIVE WEEKLY VALUE FORECAST			
Week	Day	Cumulative value	Weekly value
1	5	16 000	16 000
2	10	44 000	28 000
3	25	76 000	32 000
4	20	98 000	22 000
5	25	148 000	50 000
6	30	171 000	23 000
7	34	183 000	12 000

Figure 4.36

- *Step 1*

  List activities A to I together with details of activity durations and labour allocation (Figure 4.37).

- *Step 2*

  Prepare the precedence diagram for the sequence displaying the earliest and latest dates after completing the forward pass. This indicates an overall project duration of 29 days (Figure 4.38).

PRECEDENCE DIAGRAM ASSESSMENT OF EARLIEST START LABOUR			
Activity	Duration (Days)	Labour	
A	5	4L	Prepare an assessment
B	6	6L	of the labour
C	2	2L	requirements based
D	8	4L	on an earliest start
E	8	4L	situation
F	4	2L	
G	6	6L	
H	2	4L	
I	4	2L	

Figure 4.37

- *Step 3*
  Fully analyse the precedence diagram showing total float for each activity and then highlight the critical path (Figure 4.39).

- *Step 4*
  Prepare a labour resource histogram from a bar chart display with the labour resource allocated to each of the bar lines (Figure 4.40). The bar chart has been presented showing the critical operations first in the programmed sequence. Below the bar chart, the earliest start labour histogram has been displayed. When presenting the labour histogram, the critical labour requirements may be 'fixed' along the base of the histogram with the non-critical labour stacked on top.

- *Step 4* (alternative method)
  Prepare the labour histogram directly from the precedence diagram. This is achieved by developing the histogram based on the earliest start dates for each operation, together with respective duration and labour allocation. Why not have a go and prove how easy it really is?

Various applications of precedence techniques in relation to a range of civil engineering projects are illustrated by Cormican (1985).

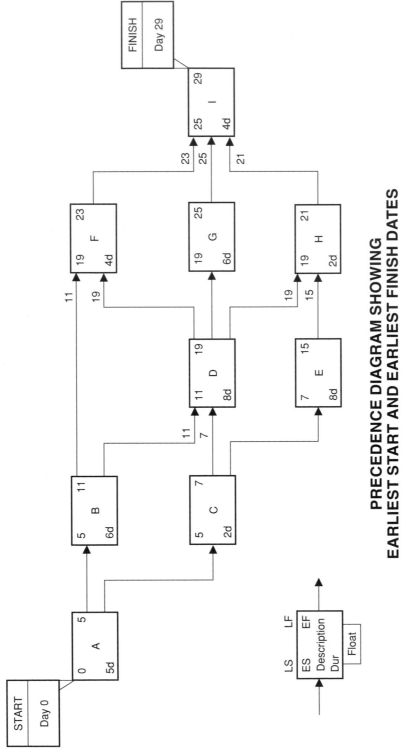

**PRECEDENCE DIAGRAM SHOWING
EARLIEST START AND EARLIEST FINISH DATES**

Figure 4.38

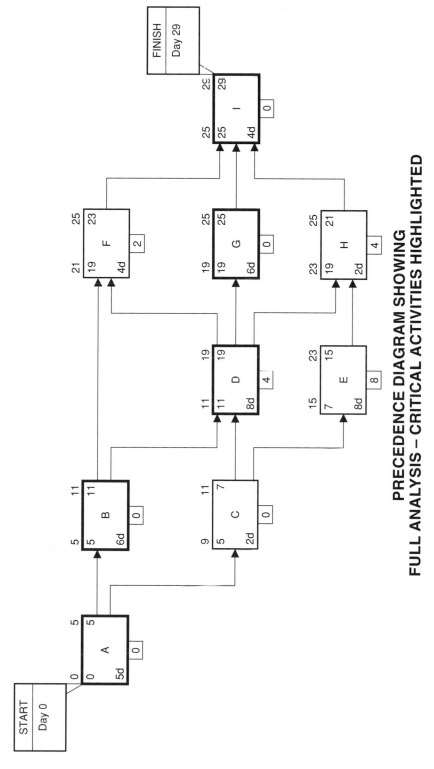

**PRECEDENCE DIAGRAM SHOWING
FULL ANALYSIS – CRITICAL ACTIVITIES HIGHLIGHTED**

Figure 4.39

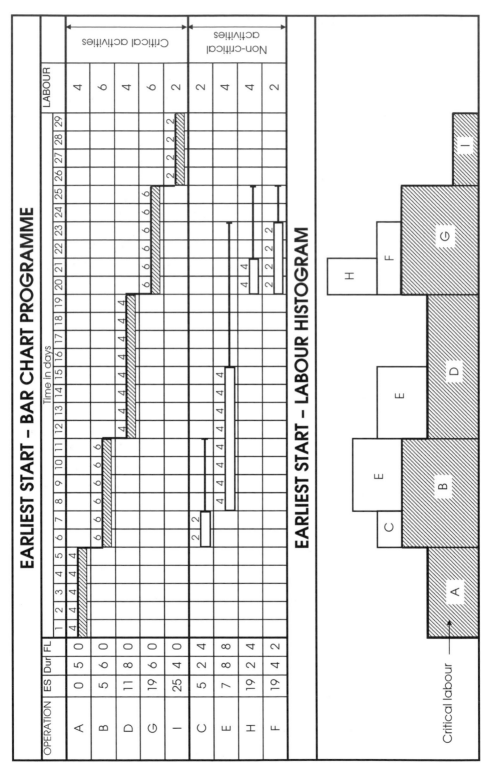

EARLIEST START – BAR CHART PROGRAMME

EARLIEST START – LABOUR HISTOGRAM

Critical labour

Figure 4.40

4.12 Line of balance (elemental trend analysis)

History and development

A line of balance diagram comprises a series of inclined lines which represent the rate of working between repetitive operations in a construction sequence. This approach was developed by the National Building Agency.

The application of line of balance to construction was pioneered by Lumsden (1965) and became recognised as the best planning method for repetitive work such as housing.

The technique has been widely used for the planning of refurbishment works, new build housing and flats and has also been applied to civil engineering works as illustrated by Harris and McCaffer (2000) and Cormican (1985). It is not unusual to see bar chart displays incorporating a line of balance diagram to illustrate the programming of any repetitive sections of the works.

Line of balance is a visual display of the rate of working of different activities on a programme. The ideal line of balance display shows all balance lines running parallel to each other, but in practice this is often difficult to achieve.

Principles of line of balance

Figure 4.41 shows the logic diagram for three operations, A, B, and C, in a construction sequence. Figure 4.42 indicates the line of balance diagram for each of these activities showing different rates of working from one operation to the other. As can be observed from the diagram, operations B and C are out of balance with operation A. However, by increasing the number of gangs employed on operations B and C the line of balance diagram as shown in Figure 4.43 can be achieved. This results in a considerable saving in the overall project period.

Figure 4.43 introduces the idea of 'buffers' at the start or finish of an operation in order to build some degree of flexibility into the programme. This also allows the preceding operation to move clear of the work area or house unit before the next operation commences.

LOGIC DIAGRAM FOR SEQUENCE

Figure 4.41

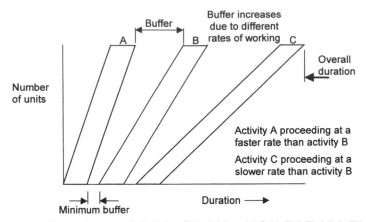

LINE OF BALANCE DIAGRAM – NON PARALLEL WORKING

Figure 4.42

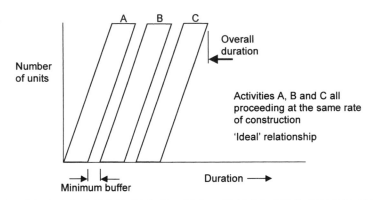

LINE OF BALANCE DIAGRAM – IDEAL RELATIONSHIP

Figure 4.43

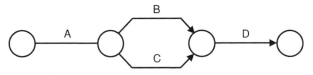

LOGIC DIAGRAM FOR SEQUENCE

Figure 4.44

Figure 4.44 shows a logic diagram for a construction sequence where operations B and C may start or finish together prior to the commencement of operation D. Figures 4.45 and 4.46 illustrate the balance diagram for these two situations depending on whether operation B is working at a slower or faster rate than operation C.

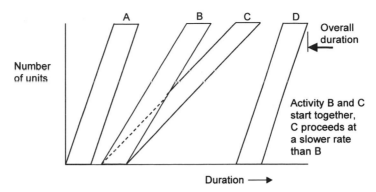

LINE OF BALANCE DIAGRAM – OPERATIONS B AND C COMMENCING TOGETHER

Figure 4.45

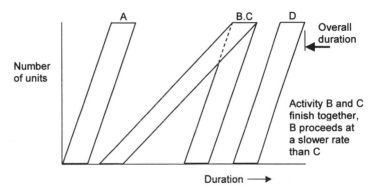

LINE OF BALANCE DIAGRAM – OPERATIONS B AND C FINISHING TOGETHER

Figure 4.46

Developing a line of balance diagram

Figure 4.47 illustrates the plan and section of a reinforced concrete elevated access platform. The work involves the construction of six bays of elevated deck together with the associated pile supports and capping beams. The sequence of the activities is shown in the logic diagram in Figure 4.48:

- Operation 1 – Install piles
- Operation 2 – In situ beams
- Operation 3 – Precast concrete deck

ACCESS DECK PROJECT

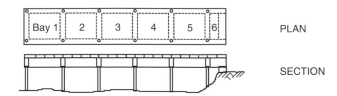

PLAN

SECTION

PLAN AND SECTION — REINFORCED CONCRETE PLATFORM

Figure 4.47

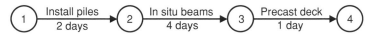

LOGIC DIAGRAM

Figure 4.48

Based on utilising one gang of men on each operation, the line of balance diagram as shown in Figure 4.49 has been developed. The planner/project manager is required to consider the most economical proposal for speeding up the project.

From the line of balance diagram it can be observed that operation 2 is out of balance with operations 1 and 3. By increasing the number of gangs on operation 2 (from one gang to two gangs) the balance lines will move into a better overall parallel position.

Figure 4.50 indicates the procedure for producing a histogram of labour resources from a line of balance diagram. This is achieved by considering the start and finish date of each of the labour gangs engaged on each operation.

Figure 4.51 shows the redrawn line of balance diagram using two labour gangs on operation 2. This results in an overall time reduction of 10 days. The revised labour resources are shown in histogram form below the line of balance diagram.

As can be seen from the above example, line of balance planning provides a visual display of the rate of working across the whole project and enables decisions to be made in relation to the use of labour. It enables the planner to start with a forecast handover rate per week and then to produce a forecast of the labour resources needed to achieve it.

The line of balance programming technique is rarely understood by construction managers and planners and is still very much underused in the construction industry. Line of balance tends to be favoured only by those who have a thorough grasp of the principles and application based on experience. Where it is company policy to use line of balance,

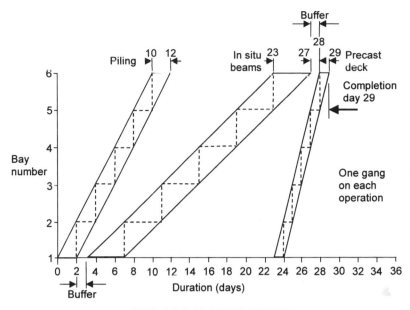

LINE OF BALANCE CHART

LINE OF BALANCE SCHEDULE – REINFORCED CONCRETE PLATFORM

Figure 4.49

it will be used, but this is in only a small number of companies. Line of balance is only really applicable to repetitive operations on refurbishment and housing projects and therefore does not have the same wide application as bar charts or networks.

Practical applications

Line of balance has gained a foothold in a number of the large housebuilding firms where the repetitive part of a construction sequence can be readily integrated into a bar chart programme.

The following example illustrates the application of line of balance to a housing project where a contractor requires a programme for the construction of ten house units. The five operations that occur in the construction sequence are shown in Table 4.2.

- *Step 1*
  Draw the sequence logic for the five operations as shown in Figure 4.52.

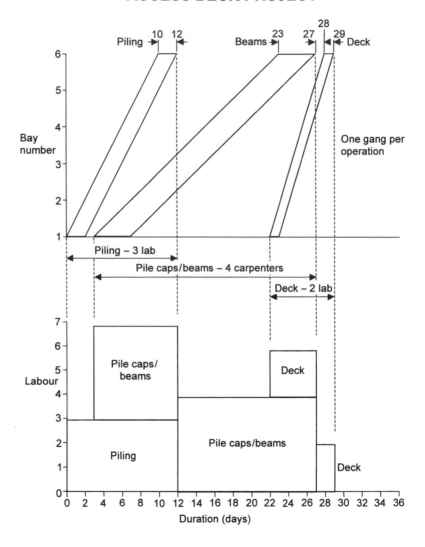

ACCESS DECK PROJECT

LABOUR RESOURCES – REINFORCED CONCRETE PLATFORM

Figure 4.50

- *Step 2*

  Assess the start and finish date of each operation in the construction sequence for the first and last units. This enables the balance lines to be plotted. Use the following formula in each case:

ACCESS DECK PROJECT

LINE OF BALANCE CHART BASED ON TWO GANGS
WORKING ON BEAMS AND CAPS

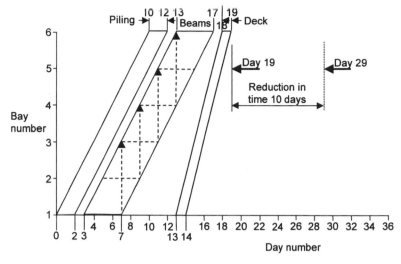

REVISED LINE OF BALANCE SCHEDULE

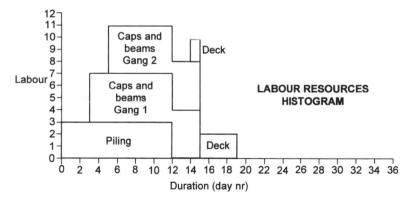

REVISED LABOUR RESOURCES

Figure 4.51

Table 4.2 Five construction operations.

Operation	Duration per unit in weeks	Number of gangs
Foundations	2	2
External walls	4	3
Roof construction	1	1
Internal finishes	4	3
External works	2	2

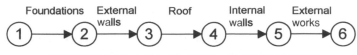

Foundations   External   Roof   Internal   External
                walls   walls   works

SEQUENCE LOGIC FOR OPERATIONS

Figure 4.52

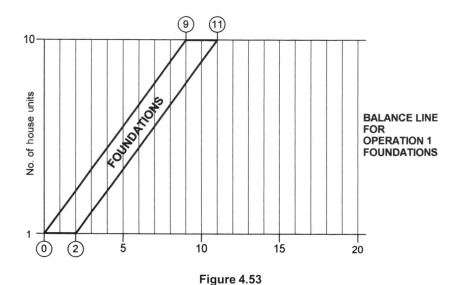

BALANCE LINE
FOR
OPERATION 1
FOUNDATIONS

Figure 4.53

$$\frac{\text{number of units less } 1 \times \text{duration of operation}}{\text{number of gangs used}}$$

Operation 1–2 (foundations)

Start of unit 1   = **week 0**

Finish of unit 1 = **week 2** (i.e. the duration of the operation)

$$\text{Start of unit 10} = 0 + \frac{(10 \text{ units} - 1) \times 2}{2 \text{ gangs}} = \frac{18}{2} = \textbf{week 9}$$

Finish of unit 10 = week 9 + 2 weeks   = **week 11**

- *Step 3*
  Plot the balance lines on squared paper. Figure 4.53 shows the balance line drawn for the foundations operation.

- *Step 4*
  Assess the start and finish date for the next operation:

Operation 2–3 (external walls)

NB Allow a minimum buffer of 1 week between one operation and the next. This will be either at the start or finish of the next operation depending on the rate of working.

Start of unit 1 $= $ week $2 + 1$ week buffer $= $ **week 3**

Finish of unit 1 $= $ week $3 + $ duration of 4 weeks $= $ **week 7**

$$\text{Start of unit 10} \quad = \text{week } 3 + \frac{(10 \text{ units} - 1) \times 4}{3 \text{ gangs}} = 3 + \frac{36}{3} = \textbf{week 15}$$

Finish of unit 10 $= $ week $15 + 4$ weeks $= $ **week 19**

Before plotting the balance line for this operation, consider the rate of construction between the foundations and external walls operations.

Foundations

$$\text{Slope of balance line} = \frac{\text{duration}}{\text{no. of gangs}} = \frac{2}{2} = 1.00$$

External walls

$$\text{Slope of balance line} = \frac{\text{duration}}{\text{no. of gangs}} = \frac{4}{3} = 1.33$$

Strictly speaking, the numbers do not really mean anything except that the lower number for the foundations operation gives an approximate indication that this operation is progressing at a faster rate than the external walls activity. This quick calculation indicates from where to plot the balance line for the external works operation. If the number is higher than that of the preceding operation, plot from the bottom of the diagram, otherwise the operations will bump into each other. If the number is lower than that of the preceding operation, plot from the top. In either case, be sure to include the minimum buffer allowance.

- *Step 5*
  Plot the balance line for external walls as shown in Figure 4.54.

- *Step 6*
  Assess the start and finish date for the next operation:

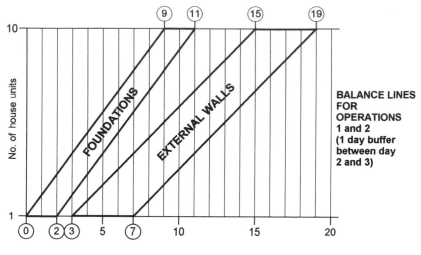

Figure 4.54

Operation 3–4 (roof construction)

Start of unit 10 = week 19 + 1 week buffer   = **week 20**

Finish of unit 10 = week 20 + 1 week duration   = **week 21**

Start of unit 1   = week $20 - \dfrac{(10-1) \times 1}{1 \text{ gang}} = 20 - 9 =$ **week 11**

Finish of unit 1   = week 11 + 1 week duration   = **week 12**

- *Step 7*
  Consider the rate of construction between the external walls and roof operations.

External walls

Slope of balance line = **1.33**

Roof construction

Slope of balance line $= \dfrac{1}{1} = \mathbf{1.00}$

As the roof operation is progressing at a faster rate than the external walls, the relationship between the balance lines will commence at the finish of the external walls operation for house unit 10.

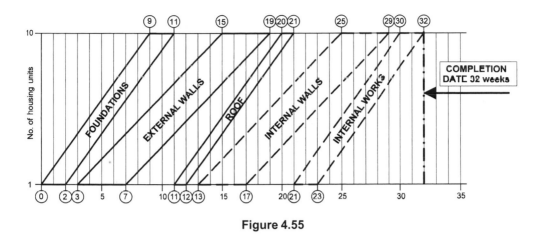

Figure 4.55

- *Step 8*
  Plot the roof construction balance line starting from the top of the diagram, as shown in Figure 4.55.

- *Step 9*
  Complete the calculations for the remaining operations. These have been added to Figure 4.55 to indicate an overall completion at week 32.

4.13 Time-chainage diagrams

History and development

In simple terms, the time-chainage diagram or location–time chart referred to by Cormican (1985) is a combination of the bar-chart and line of balance scheduling formats, and it is from these programming techniques that time-chainage principles have been developed.

Time-chainage diagrams have been widely applied on major road-works projects and in the development of the motorway system in the UK for many years. The technique was also used for the planning of tunnelling and fixed equipment installation on the Channel Tunnel project and its application is discussed in Proceedings of the Institution of Civil Engineers: The Channel Tunnel (ICE 1992). Time-chainage diagrams, like their close cousin the line of balance, are only applicable for limited types of project and therefore are not as widely appreciated in the industry as bar charts and network techniques. Nevertheless, the technique has distinct attributes and advantages on projects where it is important to depict:

- The order of activities or operations
- Where activities are happening locationally
- How activities must progress in relation to direction and distance
- Time, key dates and holidays, etc.

Principles of time-chainage diagrams

The time-chainage form of presentation enables the time dependencies between activities to be shown, together with their order and direction of progress along the job. These diagrams are most usefully employed as a planning tool on projects such as motorways and major highway works, pipelines, railway track work, tunnelling, etc.

Projects of this nature can be viewed as mainly linear in nature. In other words, construction starts at one point and proceeds in an orderly fashion towards another location. This would be typified on a highway project by activities such as fencing, drainage, road surfacing and road markings.

To some extent this type of work calls for a different planning technique because bar charts would not be useful in giving locational information, and precedence/arrow diagrams would not reflect the time–location relationship which clearly exists on such projects.

In this respect, most operations take place on a forward travel basis with the gang starting at one point or chainage and moving along the job. As one activity leaves a particular location, other activities can take its place. This ensures the correct construction sequence and avoids overintensive activity in one location.

Most of the lines on the time-chainage diagram have no appreciable thickness. This is because the time spent by each gang at a particular location is relatively small and the gang moves along the site quite quickly. Examples of this are drainage, road surfacing and safety barrier erection on a motorway.

Retaining walls would also constitute a linear activity but would tend to occupy any particular location (or chainage) for a more appreciable time due to the nature and duration of the construction operations involved.

With earthworks cut and fill operations, the situation is different in that earthworks plant will occupy a particular cut or fill zone for some time before moving to another location.

Bridges, culverts and underpasses, on the other hand, are 'static' operations and can be viewed as individual 'sites' in their own right. Such activities act as restrictions and forward travel activities may have to be programmed around them. For instance, on a highway project, drainage work may be interrupted by a bridge site and consequently the contractor will have to return later to finish the drainage once the bridge nears completion.

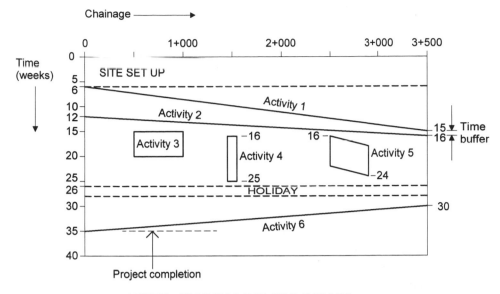

TIME-CHAINAGE DIAGRAM
Horizontal Format

Figure 4.56

Various types of time-chainage representations are possible, but basically the diagram comprises two axes, time and distance, with the various activities shown as lines or bars on the chart. Linear activities are represented with a line or bar which is positioned on the chart to show its commencing and completion chainages, and is inclined in the direction of progress at an angle consistent with the anticipated duration of the operation. A static activity, such as a bridge on a motorway, is represented by a line or thin bar positioned at a particular location or chainage, with the duration of the activity expressed by the length of the line or bar. Activity labels are annotated on the respective line or bar to distinguish one operation from another.

Figures 4.56 and 4.57 illustrate two methods of presentation of time-chainage diagrams. Chainages can either be located on the vertical or horizontal axes with time (usually in weeks) shown on the other axis.

Developing a time-chainage diagram

Drawing a time-chainage diagram is not as easy as it looks and a good deal of practice is required. Consider the following steps:

- *Step 1*
  Consult the project layout drawings and note the chainage positions. Main chainages on a highway project are at 1000 m intervals.

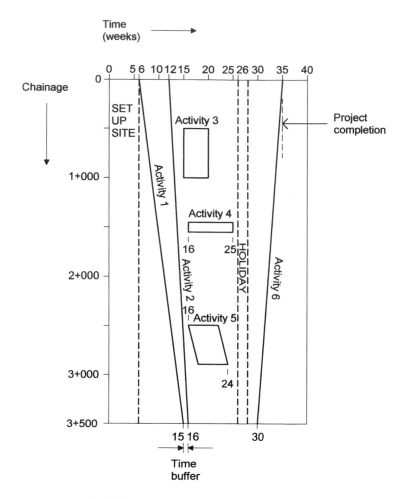

TIME-CHAINAGE DIAGRAM
Vertical Format

Figure 4.57

- *Step 2*

  Draw an outline time-chainage diagram with time along one axis and distance or chainage along the other, using either the horizontal or vertical format. Add main holiday periods allowing 2 weeks for Christmas and 1 week for Easter.

- *Step 3*

  List main programme activities or operations in approximate construction sequence. Include activities for site set-up or mobilisation and clear site at the end. Estimate the duration of each programme activity in weeks.

- *Step 4*

  Fill in the site 'set-up' and 'clear site' activities on the time-chainage diagram. Using your preferred format, plot the appropriate number of weeks over the entire length (chainage) of the project.

- *Step 5*

  Decide in turn where and when each activity will take place. For linear activities (e.g. drainage), start at the appropriate location (chainage) and week number and draw a line for the correct distance (chainage) and time (weeks). For static activities (e.g. bridgeworks), draw a thin box at the appropriate chainage with a length representing the activity duration.

- *Step 6*

  Complete all activities on the list. It is sometimes helpful to produce an outline bar chart programme to help clarify the correct time-chainage display.

Practical applications

Take, for example, a project to carry out improvements to an existing highway 3.5 km long comprising the following activities:

- Activity 1 – Fencing
- Activity 2 – Drainage
- Activity 3 – Bulk earthworks
- Activity 4 – Footbridge
- Activity 5 – Retaining wall
- Activity 6 – Road surfacing

The start of the job is at chainage 0+000 and the project finishes at chainage 3+500. A footbridge is required to cross the highway at chainage 1+500 which would be 1500 m up chainage from the beginning of the job. Distances or locations can be conveniently determined by these chainages.

This chainage referencing system is also useful on site where the contractor will usually position chainage boards along the job to enable anyone to quickly identify exactly where they are on the site.

The advantage of showing chainage along the horizontal axis is that this more readily resembles the way the drawings are laid out and the programme can therefore more easily be related to what has to be constructed. On the other hand, time-chainage diagrams showing time on the horizontal axis may be easier to read by those familiar with bar charts. The choice is a matter of personal preference.

OPERATION	TIME/LOCATION	EXAMPLE
Set up site	Starts at time 0 Duration 6 weeks	Offices for contractor and engineer
Activity 1	Starts at week 6 Duration 9 weeks Finishes week 15 Forward travel up chainage	Fencing
Activity 2	Starts week 12 Duration 4 weeks Finishes week 16 Forward travel up chainage	Drainage

Activity 3	Starts week 15
	Duration 5 weeks
	Finishes week 20
	Occupies zone between
	   chainage 0+500 and 1+000
	   for 5 weeks
		Bulk earthworks

Activity 4	Starts week 16 Duration 9 weeks Finishes week 25 Takes place at chainage 1+500	Footbridge

Activity 5	Starts week 16
	Duration 8 weeks
	Finishes week 24
	Takes place between chainage
	   2+500 and 2+900
	Forward travel up chainage at the rate
	   of 50 metres per week
		Retaining wall

Activity 6	Starts week 30 Duration 5 weeks Finishes week 35 Forward travel down chainage	Road surfacing
Holiday	Starts week 26 Duration 2 weeks Finishes week 28	Christmas shut down

TIME-CHAINAGE DIAGRAMS
Explanation

Figure 4.58

Figure 4.58 explains how the different activities are plotted on the time-chainage diagrams in Figures 4.56 and 4.57. Close inspection of Figures 4.56 and 4.57 will reveal time buffers between activities similar to those used in line of balance/elemental trend analysis.

Case Study No. 5 in Chapter 10 illustrates the practical considerations involved in developing a time-chainage diagram for a major highway project.

4.14 Accelerating the project

Project acceleration and time–cost optimisation

There are many circumstances in which the contractor may wish to speed up work on a contract, including being behind programme and having to increase production in order to minimise extensive liquidated damages. Alternatively, the client may have requested the contractor to indicate the additional costs of completing the project earlier than the contract completion date.

In practice, a department store client, for instance, may request an earlier occupation date for the building in order to take advantage of the winter or summer sales. The additional profit created by the earlier opening may well exceed the contractor's additional costs.

In order to balance the time savings against the costs of speeding up the work, optimisation studies are undertaken in order to consider the various options available. These studies allow the client and/or contractor to assess the effect on the direct and indirect costs of reducing the overall project period, and this can then be compared with the potential profits or savings in liquidated damages due to earlier completion.

Many writers refer to this method of analysis as 'time–cost optimisation', 'least cost optimisation' or 'crash costing' but perhaps a more appropriate term would be 'project acceleration' as this is really what it is all about.

Whatever you decide to call it, the first step is to understand the terminology used and then the principles of the process, illustrated here by way of two worked examples.

Project acceleration terminology

Consider the network arrow for the activity 'electrical services':

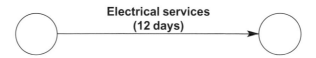

**Electrical services
(12 days)**

Normal time, normal cost

This is the usual time that would be needed to carry out the electrical services work under normal circumstances, estimated at 12 days with a cost of £10 000.

Crash time, crash cost

The crash time is the maximum time the operation can be compressed by increasing the resources. This reduction in time leads to an increase in the direct cost. The revised cost is called the crash cost. In this example, the accelerated time or crash time is to be 8 days at a total cost or crash cost of £18 000.

Cost slope

The cost slope represents the cost of accelerating any of the project activities by one unit of time (in the above case, 1 day).

In order to achieve a reduction in the overall project duration at the least possible cost, the activities on the critical path of the programme must be compressed as much as possible. This is done by first considering the activities with the least cost slope.

The cost slope of the electrical services activity is expressed as:

$$\frac{\text{increase in cost}}{\text{reduction in time}} \quad \text{or} \quad \frac{\text{crash cost less normal cost}}{\text{normal time less crash time}}$$

$$= \frac{£18\ 000 - £10\ 000}{12\ \text{days} - 8\ \text{days}} = \frac{£8000}{4\ \text{days}}$$

$$= £2000 \text{ cost slope}$$

Activity ranking

Once the arrow or precedence relationship has been analysed and the critical activities identified, each of the activities on the critical path are ranked in order of their cost slopes, starting with the least expensive.

It is obviously more economical to apply reductions in the project time to the less expensive activities first in order to achieve the required reduction in the overall project period. As the ranking is applied to the network sequence, the float times and cost slopes of non-critical activities must be considered as at some point these may become critical.

Direct costs

These are the costs associated with carrying out activities on the programme including:

- Labour
- Plant
- Materials

- Subcontractors (if applicable)
- Overheads and profit

When an activity is accelerated, the corresponding direct costs will increase. This may be due to the need to supply additional resources in the form of increased labour, plant and material requirements. Accelerating an operation may also involve overtime or weekend working. Also, incentives in the form of bonus payments may have to be made to ensure that the task is completed on time. Other direct costs may be incurred, such as additional formwork. This will reduce the number of formwork uses originally envisaged by the estimator and thus add to the direct cost of acceleration.

Acceleration may require extra direct supervision, such as foremen and gangers, to cover weekend working and the supervision of additional labour gangs.

Indirect costs

These are the time-related costs of the project which change as the project duration changes. They are normally included in the contract preliminaries and will include:

- Project supervision
- Site hutting and accomodation
- Site office telephones, heating and lighting
- Vans and site transport

When an activity is accelerated, the corresponding indirect costs will decrease. This is due to the reduction in project duration which directly affects the contract preliminaries. The contractor will, in principle, be on site for a shorter period and therefore the client will expect some reduction in the site administration costs or preliminaries.

Total project cost

This is the summation of the direct and indirect costs. The total cost is usually expressed at the normal time and at the optimum project duration.

Optimum project duration

The optimum project duration occurs at the point where the most beneficial least cost situation occurs, taking into account both direct and indirect costs. In order to establish the least cost situation the cost increase for each unit of time reduction must be considered.

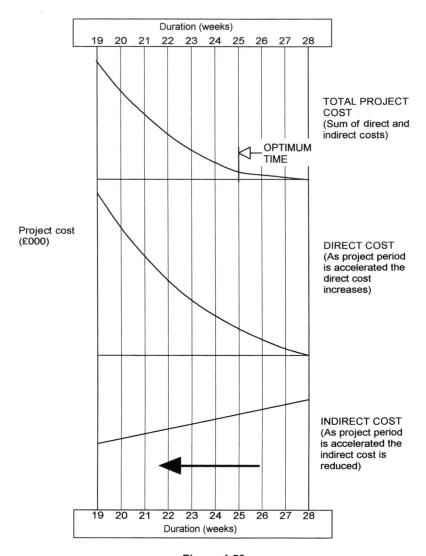

Figure 4.59

Figure 4.59 illustrates the relationship between direct cost, indirect cost and project duration. The summation of the direct and indirect costs (i.e. the total project cost) is also indicated. The optimum duration is shown on the total cost graph and is the date at which the costs rise most significantly.

Project acceleration principles

Application 1 – based on a network arrow diagram

Figure 4.60 shows a construction sequence involving activities A to I in arrow diagram format. The tabular data indicates the normal times, normal costs, crash times and crash costs. The indirect costs (or time-related preliminaries) of the project amount to £2000 per week.

Figure 4.61 illustrates the analysed arrow diagram based on the normal time situation which gives an overall project duration of

INITIAL ARROW DIAGRAM

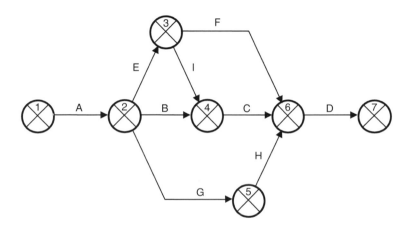

PROJECT DATA				
Activity	Normal time	Normal cost	Crash time	Crash cost
A	6	12 000	4	20 000
B	8	24 000	4	48 000
C	3	18 000	2	20 000
D	6	18 000	4	24 000
E	6	36 000	4	54 000
F	10	10 000	6	50 000
G	5	20 000	3	30 000
H	8	40 000	6	50 000
I	2	20 000	2	20 000
Summation	£198 000			
Indirect costs £2 000 per week				

Figure 4.60

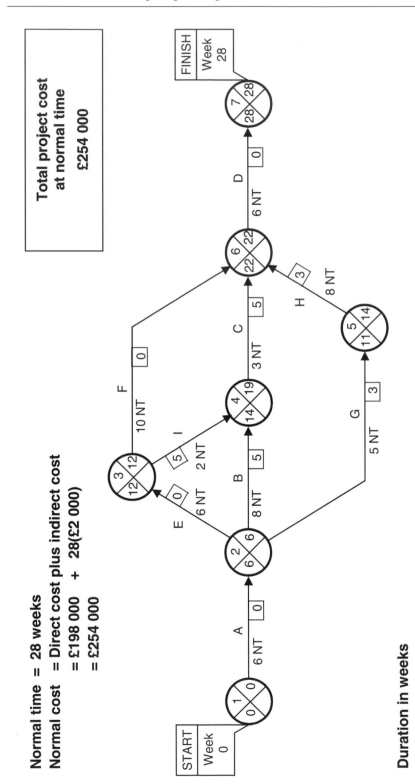

**Total project cost
at normal time
£254 000**

Normal time = 28 weeks
Normal cost = Direct cost plus indirect cost
= £198 000 + 28(£2 000)
= £254 000

Duration in weeks

NORMAL TIME/NORMAL COST ANALYSIS

Figure 4.61

28 weeks. The total project cost, based on normal time, has been calculated at £254 000 which is made up of direct costs of £198 000, plus £56 000 (28 weeks × £2000) of indirect costs.

Two scenarios will now be considered.

Scenario 1 – a reduction of 5 weeks in the project duration in order to complete the project by week 23. The effect on the direct and indirect costs will be considered in order to achieve this reduction in time.

In order to reduce the project duration by 5 weeks, it will be necessary to consider the cost slopes of all activities. The assessment of the cost slopes and their appropriate ranking are indicated in Figure 4.62.

From Figures 4.61 and 4.62 it can be seen that in order to reduce the project duration by 5 weeks, it will be necessary to reduce the duration of activities on the critical path in rank order (from the least expensive to the more expensive cost slopes). In Figure 4.61, the critical path follows activities D, A and E in rank order. By using the crash durations for these activities, a 5-week reduction in the overall project period can be achieved, as shown in Table 4.3.

Therefore, in order to achieve an acceleration of 5 weeks, the project cost will be increased by £13 000, i.e.:

Crash cost   – £267 000
Less normal cost = £254 000
Acceleration cost =   **£13 000**

Figure 4.63 shows the revised arrow diagram analysis using the crash times on activities A, E and D. This analysis also indicates the revised float times on the non-critical activities.

Table 4.3 Reduction in project period.

Activity	Normal time (weeks)	Crash time (weeks)	Reduction in time (weeks)	Cost slope (£)	Increase in direct cost (£)
D	6	4	2	3 000	6 000
A	6	4	2	4 000	8 000
E	6	5	1	9 000	9 000
Reduction in overall time =			**5**		
Increase in direct cost due to this reduction in time =					**23 000**

Table 4.4 Total project costs at week 23.

Direct costs	=	£198 000 + £23 000 =	£221 000
Indirect costs	=	23 weeks @ £2 000 =	£46 000
Total project cost =			**£267 000**

COST SLOPE ASSESSMENT

Activity	Normal time	Crash time	Saving in time	Normal cost ($000)	Crash cost ($000)	Increase in cost ($000)	Cost slope	Float	Order/ Ranking
A	6	4	2	12	20	8	4000	Zero	2nd
E	6	4	2	36	54	18	9000	Zero	3rd
F	10	6	4	10	50	40	10000	Zero	4th
D	6	4	2	18	24	6	3000	Zero	1st
B	8	4	4	24	48	24	6000	5	Non critical operations
G	5	3	2	20	30	10	5000	3	
H	8	6	2	40	50	10	5000	3	
I	2	2	0	20	20	0	0	5	
C	3	2	1	18	20	2	2000	5	

Figure 4.62

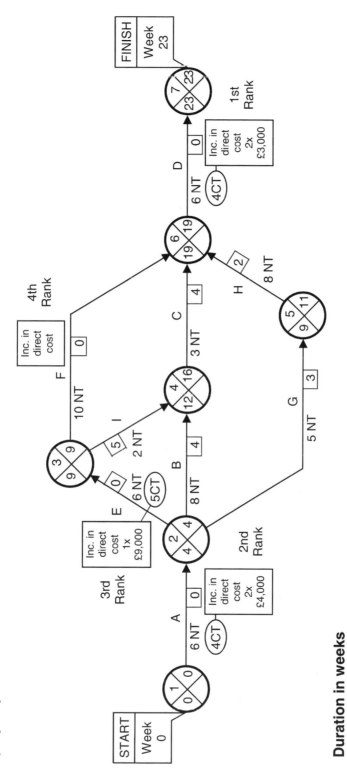

**Analysis based on accelerated
project period of 23 weeks**

REVISED ARROW DIAGRAM

Figure 4.63

Duration in weeks

Week No. start	Direct cost	+ve	Indirect cost	−ve	Aggregate	Total project cost (£000)	Activity ranking
28	198		56			254	
27		+3		−2	+1	255	Activity D
26		+3		−2	+1	256	Activity D
25		+4		−2	+2	258	Activity A
24		+4		−2	+2	260	Activity A
23		+9		−2	+7	267	Activity E
22		+9		−2	+7	274	Activity E
21		+10		−2	+8	282	Activity F
20		+10		−2	+8	290	Activity F
19		+10		−2	+8	298	Activity F
18		+10		−2	+8	306	Activity F

CALCULATION OF CHANGE IN DIRECT AND INDIRECT COST PER WEEK AFTER APPLYING RANKING

Figure 4.64

Scenario 2 – the effect on the project cost for each week's reduction in time, in order to assess the least cost situation.

The change in the direct and indirect costs due to each week's reduction in the project time is indicated in Figure 4.64. This relationship is presented in graphical form in Figure 4.65.

The relationship between the indirect and direct costs for each week of the project from weeks 28 to 18 can be observed in Figure 4.65. The summation of the direct and indirect costs is displayed and the point on the graph where the cost suddenly increases is the position of the optimum time and least cost situation. This occurs at the end of week 24. At week 25 the change in cost per week alters from £2000 to £7000 per week.

INDIRECT COST, DIRECT COST AND TOTAL PROJECT COST RELATIONSHIP

GRAPHICAL PRESENTATION

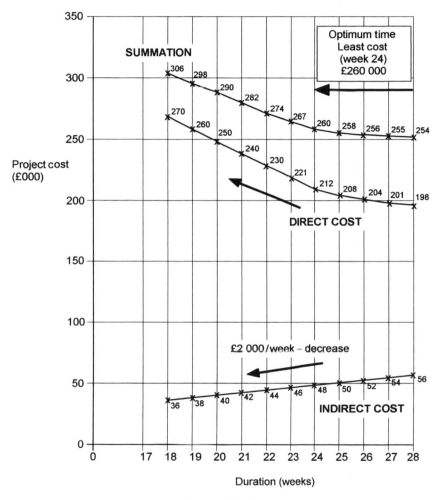

Figure 4.65

The project costs have been analysed back to week 18 in order to provide an overall assessment of the project cost situation.

Application 2 – based on a Precedence Network

Figure 4.66 shows a precedence diagram for activities A to J during a construction sequence. Details of the normal time, normal cost, crash time and crash cost are given in Table 4.5.

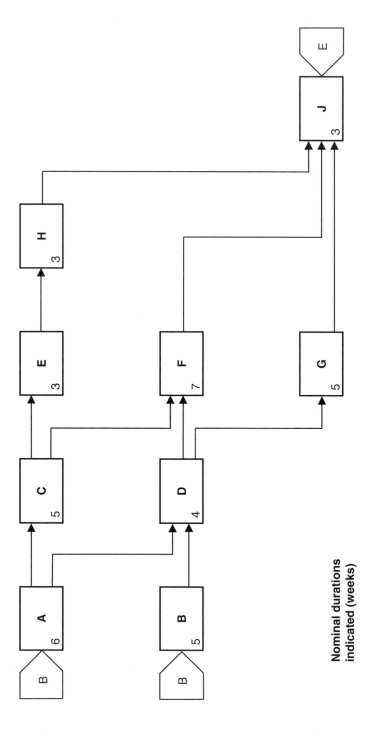

Nominal durations indicated (weeks)

INITIAL PRECEDENCE DIAGRAM

Figure 4.66

Table 4.5 Time and cost for construction activities.

Activity	Normal time (days)	Normal cost (£)	Crash time (days)	Crash cost (£)
A	6	18 000	3	21 000
B	5	14 000	3	18 000
C	5	16 000	4	18 000
D	4	8 000	4	8 000
E	3	10 000	3	10 000
F	7	22 000	5	28 000
G	5	4 000	4	5 000
H	3	12 000	2	13 000
J	3	10 000	2	15 000
Total normal cost	=	**114 000**		
Indirect cost (preliminaries)	=	**1000 per week**		

Table 4.6 Cost slope of activities.

Activity	Time (weeks) Normal	Crash	Saving	Cost (£000) Normal	Crash	Increase	Cost slope £000	Float	Rank
A	6	3	3	18	21	3	1	C	1
C	5	4	1	16	18	2	2	C	2
F	7	5	2	22	28	6	3	C	3
J	3	2	1	10	15	5	5	C	4
B	5	3	2	14	18	4	2	2	
D	4	4	–	8	8	–	–	1	
G	5	4	1	4	5	1	1	3	
E	3	3	–	10	10	–	–	1	
H	3	2	1	12	13	1	1	1	

C = critical

An assessment of the optimum time and optimum cost solution for the project is required in order to complete the construction sequence as economically as possible. Figure 4.67 shows the analysis of the precedence diagram based on normal time. The overall project period is 21 weeks and the critical path follows the activities A, C, F and J. The cost slope for each activity may be assessed in tabular form in order to simplify the analysis. The activities are listed in Table 4.6 in critical order, followed by non-critical activities in earliest start order.

The purpose of the analysis is to consider the effect on direct cost of reducing the overall duration 1 week at a time. The activities should be considered in the order of their cost slope values and ranking. Figure 4.68 shows the revised precedence diagram analysis using the new durations, which enable the overall project period to be reduced to 16 weeks.

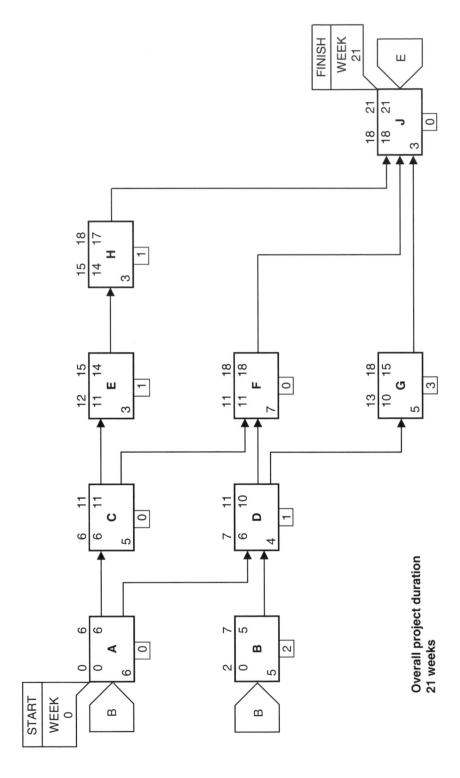

ANALYSIS OF NORMAL TIME

Figure 4.67

Overall project duration
21 weeks

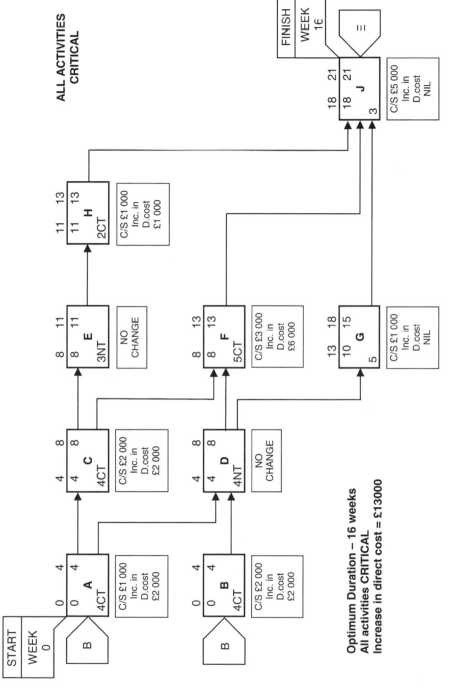

ALL ACTIVITIES CRITICAL

START	WEEK 0

0	4
0	4
A	
4CT	

C/S £1 000
Inc. in
D.cost
£2 000

0	4
0	4
B	
4CT	

C/S £2 000
Inc. in
D.cost
£2 000

4	8
4	8
C	
4CT	

C/S £2 000
Inc. in
D.cost
£2 000

4	8
4	8
D	
4NT	

NO
CHANGE

8	11
8	11
E	
3NT	

NO
CHANGE

8	13
8	13
F	
5CT	

C/S £3 000
Inc. in
D.cost
£6 000

13	18
10	15
G	
5	

C/S £1 000
Inc. in
D.cost
NIL

11	13
11	13
H	
2CT	

C/S £1 000
Inc. in
D.cost
£1 000

FINISH	WEEK 16

18	21
18	21
J	
3	

C/S £5 000
Inc. in
D.cost
NIL

Optimum Duration – 16 weeks
All activities CRITICAL
Increase in direct cost = £13000

RE-ANALYSIS TO OBTAIN OPTIMUM DURATION

Figure 4.68

Table 4.7 Change in direct costs.

Activity	Reduction in time (weeks)	Cost slope	Critical or non-critical	Rank	Increase in direct cost
A	2	£1000	C	1	£2000
B	1	£2000	NC		£2000
C	1	£2000	C	2	£2000
D	no change	—	—		—
E	no change	—	—		—
F	2	£3000	C	3	£6000
G	no change	—	—		—
H	1	£1000	NC		£1000
J	no change	—	—		—
	= 5 days reduction in project duration				
	Total increase in direct cost =				**£13 000**

Total project cost at week 16
Direct cost at week 21	= £114 000
Additional direct cost	= **£ 13 000**
Total direct cost	= £127 000
Indirect cost	
= 16 weeks @ £1000	= £ 16 000
Total project cost at week 16	**= £143 000**

Notes have been included in Figure 4.68 to indicate the increase in direct cost attributed to each activity.

The analysis has resulted in all the activities becoming critical. This is a common situation which occurs when using project acceleration or time–cost optimisation techniques. A summary of the effect on the direct cost of using the revised durations is indicated in Table 4.7.

References

CIOB (2002) *Chartered Institute of Building Code of Practice for Project Management*, 3rd edn. Blackwell Publishing.

Cormican, D. (1985) *Construction Management: Planning and Finance*. Construction Press.

Harris, F. & McCaffer, R. (2000) *Modern Construction Management*, 5th edn. Blackwell Publishing.

ICE (1992) Proceedings of the Institution of Civil Engineers. *The Channel Tunnel Part 1: Tunnels*. Thomas Telford.

Illingworth, J. (1993) *Construction Methods and Planning*. E. & F. N. Spon.

Kempner, T. (1980) *A Handbook of Management*, 3rd edn. Penguin Books.

Lumsden, P. (1965) *Programming House Building by Line of Balance*. National Building Agency.

Nuttall, J. F. & Jeans, R. E. (1960a) The critical path method. *BRE Current Papers, Construction Series 3.*

Nuttall, J. F. & Jeans, R. E. (1960b) CPM applied to building site control. *BRE Current Papers, Construction Series 12.*

Reece, G. (1989) The future of project planning. *Construction Manager*, Journal of the Chartered Institute of Building, May.

5 Pre-tender planning

5.1 Introduction

The pre-tender planning process starts well before the contractor is invited to tender for a contract. Construction companies are in business like those in any other industry and competition for work is fierce. Consequently contractors need to market their skills and convince prospective customers that they can offer the range of services required to satisfy the client's needs and expectations.

A top-ten contractor will typically enjoy a market share of 2–3% of the £65 billion annual output of the industry and the largest of these will undertake 6–7% of the total work available. To win this level of work requires a great deal of effort and therefore the contractor needs to be alive to business opportunities by checking local planning registers, reading the *Official Journal of the European Community* (OJEC) and being included on the tender lists of all the major clients, consultants, local authorities and central government departments and agencies.

The pre-tender planning process is described below under the following headings:

- Establishing contract leads
- Tender pre-qualification
- The tendering process
- Traditional competitive tendering

5.2 Establishing contract leads

A marketing approach

All construction firms need to develop contacts in order to secure opportunities to tender for work or to be in the right place at the right time when the chance to negotiate contracts presents itself. The opportunity to tender for a factory extension for Acme Widgets is more likely to arise in the private boxes at Old Trafford or the Cheltenham Festival than in a back street public house bar!

Within medium and large-sized companies, however, a formalised marketing approach is common so that contract leads can be developed strategically at an early stage so as not to miss potential opportunities to be considered for inclusion on the final tender list.

Tendering costs money and therefore contractors cannot afford to tender for contracts willy-nilly without any thought as to the risk and profit potential or whether the type of work or contractual arrangements are suitable.

It is clearly important then for the contractor to be on the tender lists for jobs he is keen to win and not to waste time and money tendering or taking 'cover prices' for unattractive contracts. Indeed, a more selective tendering strategy may increase the contractor's tender success rate from say 1 in 8 bids to 1 in 6 or better.

Public sector projects

In the public sector, major construction works or supply contracts are advertised in the OJEC. This publication covers tendering opportunities over £300 000 in the public and government sectors for projects such as motorways, hospitals, community buildings and projects involving European funding.

These advertisements follow a standard procedure and public works projects and supply contracts are notified in the local language under standard headings. A typical tender opportunity for a hospital project in Sunderland would include the following list of information:

- Awarding authority
- Award procedure and contract type
- Site location and design information available
- Completion deadline
- Legal arrangements for joint venture or consortia bidders
- Deadline, address and language for submitting tender list applications
- Final date for dispatch of tender invitations
- Deposits, bonds or guarantees required
- Financing and payment (e.g. monthly payment against invoice)
- Qualifications such as past experience, technical expertise and financial stability
- Award criteria (e.g. price, technical merit, quality, value for money)
- Variants, which would include permissibility of tender qualifications or alternative bids
- Other information about the project in hand
- Various official dates and notices

Due to the extensive numbers of advertisements and the language difficulty, most contractors use the services of the local Euro Information Centre or perhaps a firm of contract leads consultants in order to find out what is happening in the market place. A selective approach to following up potential leads can be undertaken through these sources based on searches using, say, five preferred criteria, for example:

- Project type
- Project size
- Contractual arrangements
- Form of contract and contract terms
- Location

Private sector projects

Contract leads may also be obtained from other sources:

- Local authority planning applications
- Planning committee meetings
- Minutes of local authority meetings
- Public press announcements
- Trade journals and magazines
- Trades representatives (they are often 'in-the-know')
- Keeping your eyes and ears open (don't forget Old Trafford!)

Whatever the source, a lead can take a considerable time to come to fruition. For instance, an average contract lead can take 10 months to track from an initial planning application to inclusion on the tender list.

5.3 Tender pre-qualification

More often than not, contractors are required to undergo some form of pre-qualification in order to be selected for inclusion on a tender list. This is a formal process which usually requires prospective tenderers to answer a standard questionnaire and perhaps attend a formal interview and make a presentation.

On most projects, the main contractor will usually be appointed as the principal contractor under the Construction (Design and Management) Regulations 1994, and therefore pre-qualification will also include questions about their health and safety record, health and safety training and the qualifications and experience of their staff and operatives. In all cases, contractors must be able to demonstrate that they have appropriate procedures in place to comply with health and safety law, as well as possessing the usual qualities and resources expected of a competent contractor.

Consideration of the contractor's financial standing would involve scrutiny of the contractor's accounts to determine profitability, solvency, liquidity, asset strength and payment record, as well as taking out references from previous clients, bankers and the trade, e.g. suppliers and builders' merchants.

Where sensitive information is concerned, trade and other references may be subject to the Data Protection Act 1998. For instance, normal trade information concerning a contractor's financial stability, the length of time taken to settle invoices or the contractor's credit limit with suppliers and builders' merchants might be caught by the Act.

The legislation protects individuals from the improper use and dissemination of sensitive data by data controllers but, whilst the Act does not apply to incorporated companies, little distinction seems to be made in practice between sole traders, partnerships and limited companies.

In a limited survey conducted by the authors:

- Local authorities were reluctant to provide financial references for contractors on their approved lists
- Some materials suppliers were cautious about the sort of 'sensitive' information they were prepared to give
- Some firms of quantity surveyors and architects establish a contractor's financial standing by asking how much they pay for providing contract bonds or insurances, some carry out company searches through Companies House and some use credit reference agencies which have special status under the Act

In any event it seems that the days of the 'trade reference' may be numbered.

As the construction industry moves towards a less adversarial approach to contracting, a factor of particular importance to some clients is the contractor's attitude towards claims. This may be seen as reflecting their ability to engage in partnering and more open methods of doing business.

Examples of issues which should be considered when choosing prospective tenderers are as follows.

General criteria

- Company details
- Completed contract record
- Technical expertise
- Plant and resources
- Financial accounts
- References

Competence

- Past experience
- Response to health and safety plan
- Organisation
- Safety advice
- Training provision
- Management procedures
- Accident statistics and prosecutions
- Vetting and control of subcontractors
- Safety management system

Resources

- Provision for health and safety
- Plant and equipment
- Technical support
- Trained personnel
- Adequate time

As well as the general information listed above, pre-qualification may also involve submitting a company video and giving a formal Powerpoint-type presentation. Many companies produce formal pre-qualification documents or brochures, the typical contents of which might be:

- Company organisation and structure
- Regional/national/international offices
- Key statistics, e.g. registered office, bankers, capital structure, annual turnover, profit progression, 5-year summary of accounts
- Audited annual accounts for the last 3 years
- Current projects
- Relevant completed projects
- Special expertise or experience
- Résumés of key personnel
- Project organisation structures and management approach
- Typical outline method statement
- Typical outline programme
- Cost planning approach for management or target contracts
- Details of insurances
- Evidence of quality assurance certification
- Health and safety policy and safety record statistics

Company brochures can be produced impressively and professionally in-house using modern computer software such as desk-top publishing packages and photo-image software for presenting completed projects in colour. Project management software such as Power Project can be used to present typical contract programmes in colour format.

An average cost of around £500 per brochure is a relatively small investment for the opportunity to be included on a tender list for an £8 million warehouse project.

5.4 The tendering process

The pre-tender planning process begins with the tender enquiry to the contractor and ends with the contractor's tender submission. The steps in the process vary according to the procurement method chosen by the client or his advisors. Figure 4.2 (Chapter 4) illustrates the stages involved.

A central feature of pre-tender planning is the tendering process which is described in detail in the CIOB Code of Estimating Practice (CIOB 1997). This provides an authoritative guide to good practice and is based on procedures developed by a wide range of construction organisations. The Code distinguishes between estimating and tendering, which are separate though closely interrelated processes.

The estimating stage of a bid is simply putting together the numbers. It is the 'technical' process of arriving at the cost of a proposed project by taking into account all factors that contribute to the costs of construction. Estimating involves:

- Making enquiries to suppliers for the cost of materials
- Asking for specialist quotations from subcontractors
- Arriving at the cost of employing different types of labour, including the on-costs such as national insurance and holidays with pay contributions
- Pricing the items in the bills of quantities provided in the tender enquiry from the client
- Planning the time-scale for the project (usually with other colleagues)
- Pricing the cost of running the project on site (preliminaries)
- Producing a summary of the 'tender' for directors

Tendering, on the other hand, is converting the numbers into a competitive bid after consideration of market factors and risk. It is carried out by the directors and the senior managers of the company, taking into account such factors as who the client is, the level of competition at the time of tendering, the technical and commercial risks involved and whether the company needs the work or not. This is essentially a commercial decision which may result in making alterations to the estimate by adding or deducting sums from the total arrived at by the estimator(s). A decision is then made on the final figure to be submitted to the client.

The CIOB Code contains many interesting and practice-based proformas which a contractor may consider using to collate data and build up his estimate. A series of flow charts illustrates the various stages of the estimating process. Further explanation, together with worked examples, can be found in a number of textbooks including Brook (1993), Smith (1995) and Bentley (1987), as well as various CIOB Construction Papers, (CIOB undated).

The following description of pre-tender planning procedures is typical of those used by many medium and large-sized contractors. Procedures for a traditional competitive tender are described first, followed by a comparative description of procedures for a design and build tender.

5.5 Procedures for a traditional competitive tender

The procedures outlined below are those undertaken within the medium/ large construction organisation. Within larger firms a more structured functional approach to the estimating and tendering process is often developed. These approaches are illustrated in Figures 5.1 and 5.2.

In either case, the contractor will have a lot to consider during the preparation of the estimate and its subsequent conversion into a tender bid submission. The following stages summarise the pre-tender planning sequence:

- Decision to tender
- Pre-tender arrangements

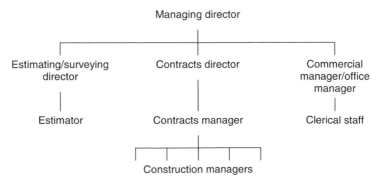

SMALL/MEDIUM SIZED COMPANY ORGANISATION STRUCTURE

Managing director

Estimating/surveying director   Contracts director   Commercial manager/office manager

Estimator   Contracts manager   Clerical staff

Construction managers

Figure 5.1

LARGE COMPANY – FUNCTIONAL ORGANISATION STRUCTURE

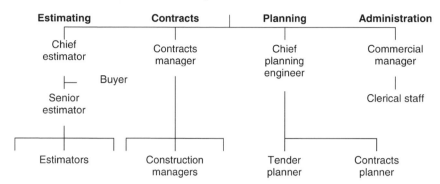

Figure 5.2

- Site visit and report
- Enquiries to subcontractors and suppliers
- Tender method statement
- Build-up of estimate
- Pre-tender programme assessment
- Build-up of contract preliminaries
- Response to pre-tender health and safety plan
- Management adjudication and risk assessment
- Analysis of tender performance

Within the medium-sized company it is likely that there would be no separate buying facility available and that the company would not have any specialist planning staff to call upon. Responsibility for estimating and tendering would lie with three principal directors. Their involvement in the bidding process is shown in Table 5.1.

Within larger organisations, a full tender team would be involved in the tender planning process. This would be based on a functional organisation structure from which a dedicated estimating team would be developed. The team would have full access to:

- A buying section to deal with subcontractor and material suppliers' enquiries
- A planning service usually provided by the planning department
- Full administration support staff for dealing with tender enquiries and correspondence
- Senior construction personnel for advice on methods of construction, plant and preliminaries

Table 5.1 Principal directors' tendering responsibilities.

Personnel	Involvement
Managing director	**Decision to tender** Organisation of bid Monitoring progress **Tender adjudication and risk assessment** **Analysis of tender performance**
Estimating/surveying director	**Decision to tender** Site visit and report Enquiries to subcontractors and suppliers Tender method statement Build-up of estimate Build-up of contract preliminaries **Tender adjudication and risk assessment** **Analysis of tender performance**
Contracts manager/director	**Decision to tender** Site visit (with estimator) Advice on construction methods Advice on plant and preliminaries Pre-tender programme Response to pre-tender health and safety plan **Tender adjudication and risk assessment** **Analysis of tender performance**

Figure 5.3 shows a programme of tender activities for a traditional competitive tender, highlighting the involvement of key personnel within the overall estimating and tendering period. A minimum period of 4 weeks is normally allowed from tender enquiry to tender submission.

5.6 Procedures for a design and build tender

With a design and build arrangement, the contract may be based on either a partial or full contractor's design. In either case, the tendering contractor will need time to arrange for design work to be prepared in sufficient detail to enable quantities to be prepared for pricing purposes. This is an added risk factor for the contractor. The overall stages involved in pre-tender planning will vary according to the extent of design input by the client or contractor at the tender stage.

The pre-tender planning stage will involve input from either the contractor's in-house design team or independent design consultants appointed by the contractor. With many design and build projects, extensive liaison with subcontractors forms a necessary part of the tendering process.

PROGRAMME OF TENDER ACTIVITIES (Traditional contract)

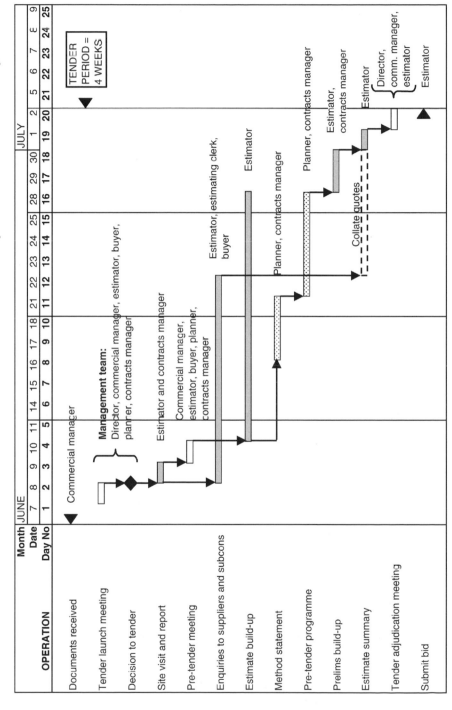

Figure 5.3

In the larger organisations, a bid manager (senior estimator) may be appointed on major tenders to coordinate information flow during the estimate preparation and tender stage. The role of the bid manager would involve responsibility for:

- Receipt of tender documentation
- Attendance at the tender clarification meeting
- Organisation and distribution of documents and correspondence received during the tender
- Managing the tender launch and decision to tender meeting
- Updating and reporting to senior management on the progress position during the tender preparation
- Attending tender adjudication meeting and any settlement meetings during estimate
- Arranging for bid submission on appropriate date

Figure 5.4 shows the various stages involved in the organisation of a design and build tender in bar chart form.

Project success often relies on the ties developed at tender stage between the subcontractors and the main contractor, and many of the subcontract packages will involve responsibility for design. The design and build contractor often has only one chance of getting it right and whether or not he does largely depends on his ability to manage his subcontractors during the design as well as the construction period of the project. Trust between both parties is essential for a successful project.

Figure 5.4 identifies the design and build contractor's responsibility for both quantity assessment and design and also shows the requirements for close liaison with work package subcontractors during the bid preparation.

5.7 Traditional competitive tendering

Competitive tendering is still the most widely used method of procurement in construction despite the popularity of design and build, construction management and partnering (see Chapter 1). Irrespective of the procurement method used, however, the principles which apply to traditional tendering are generally applicable as in most cases the client will require a price for the job and the contractor will have to be sure that he is giving the right price to enable the contract to make a profit.

The remainder of this chapter is devoted to the procedures inherent in preparing a 'traditional' tender, which comprise the following stages:

- Decision to tender
- Pre-tender arrangements

PROGRAMME OF TENDER ACTIVITIES (Design and build contract)

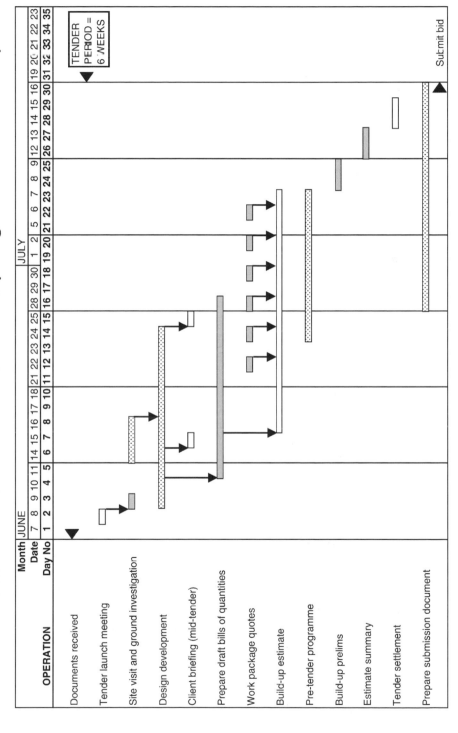

Figure 5.4

- Site visit report
- Tender enquiries to subcontractors and suppliers
- Build-up of the estimate
- Pre-tender method statement
- Preparation of pre-tender programme
- Build-up of contract preliminaries
- Response to pre-tender health and safety plan
- Management adjudication of the estimate
- Analysis of tender performance

5.8 Decision to tender

The decision to tender needs to be carefully considered by the contractor as tendering costs money. Several factors have to be weighed up, as listed below. Not least of these factors is the 'quality' of the enquiry both in terms of the status of the client and the prestige of the project.

Some contractors grade the tender enquiries received from clients at the decision to tender stage. Status grades from 1–4 may be applied. However, there appears to be little to be gained from this practice especially in terms of how much effort the contractor puts into a grade 2 tender as opposed to one ranked at grade 4.

A more sensible approach might be for contractors to consider each tender on the basis of submitting a winning bid, in which case the ultimate price should be influenced only by risk and market factors. Where circumstances change during the bid process – perhaps the contractor wins another contract unexpectedly – this can be taken into account at the adjudication stage.

A company's tender success rate may be influenced more by being selective at the decision to tender stage than by developing some esoteric tender grading system.

Check-list

The following factors will influence tendering decisions in general, together with specific market conditions at the time of tendering.

General tendering policy

- Is it our kind of work – does it fit into our strategic plan?
- What is the current workload in both the contracts division and estimating section?
- Do we have the financial and management resources to undertake the work?

Working capital

- Is sufficient working capital available to fund the project?
- What will be the effect on company financial resources?
- The working capital required to fund a £500 000 project will be approximately 15–20% of the monetary value at the peak funding month (say £100 000–£150 000)

Availability of resources

- General management personnel (e.g. contracts managers, planning engineers, quantity surveyors)
- Site management (e.g. site agents, foremen/gangers, site engineers)
- Labour and plant
- Subcontractors – are suitable subcontractors available and what is their resource situation?

Location

- Is the project located within our trading area?
- What management and control problems will there be with a contract located some miles from head office?

Size and type of work

- What is the monetary value of the project?
- Is the contract too big for the company to undertake?
- Taking on a project which is too big could be damaging to future planning and growth
- What impact will there be on the viability of the business if the contract fails to make an adequate margin?
- If a contractor with an annual turnover of £10 million wins a £4 million contract and this project makes a loss, the whole business could be put at risk
- A major project could give the company severe liquidity problems
- How did the company perform on similar types of work in the past?

Subcontract element

- What is the extent and value of the contractor's work in the project compared with the subcontract element?
- Is the main contractor simply being asked to manage a number of subcontractors?
- Is a reasonable mark-up on subcontractors likely?

- Is there a risk of incurring liabilities from subcontractor claims?
- What future liabilities may be incurred as a result of defective work by subcontractors?

Degree of competition

- How many contractors have been invited to tender?
- Which contractors are on the tender list?
- Obtain details of competitors by making contact with material suppliers or specialist subcontractors listed in the contract documents
- Do we want the work or should we take a cover price (i.e. an arrangement whereby one contractor is given a price by another contractor which is then submitted as a tender offer. The price will be sufficiently realistic to look like a bona-fide tender but high enough so as not to win the contract)

Tender period

- Have the recommendations of the National Joint Consultative Committee (NJCC) Code of Procedure for Selective Tendering been followed?
- For a contract value of £500 000, a period of 4–6 weeks should be allowed for tendering

Terms and conditions of contract

- Which form of contract is to be used?
- Is there any contractor design element?
- Are there any amendments to the standard form?
- Is the contract period stated in the documents?
- What is the rate of liquidated and ascertained damages?
- Is retention applicable?
- What is the defects liability period?
- Is a bond to be provided by the contractor?
- What are the insurance requirements?
- What are the payment conditions and terms?

Tender documentation

- Is there any quantity risk placed on the contractor (e.g. drawings and specification tender)?
- What is the quality and accuracy of the documents provided?
- Is there potential for contractual risk in the contract documentation (e.g. liability for ground conditions)

- If the tender documentation is poor, what problems or potential claims could arise?
- Potential for claims is not always attractive to contractors as there is no guarantee of success, large sums of money can be outstanding for long periods and the costs of compiling claims can be high

Client and other participants

- What is the name of the client?
- What is the name of the architect, engineer, quantity surveyor, etc?
- Have there been any previous difficult experiences working with these people?
- Do we need to allow a contingency factor for this?
- What is the financial standing of the client?

Market factors

- What is Government policy (now and future) and its effect on the construction sector?
- What are the general market conditions in relation to the availability of work?
- What is the current level of bank interest rates?
- Is our current borrowing facility adequate?
- Is there a feel-good factor in the business economy?

The decision to tender is the responsibility of senior management. In order to aid this decision, project particulars may be set out on a Preliminary/Tender Enquiry Form as shown in the CIOB Code of Estimating Practice (1997).

Tendering costs

From a limited survey of contractors, Hughes (2004) has found that contractors can spend in the order of 3% of turnover on winning work. This includes not only tendering costs but also the costs of marketing and following up enquiries.

Bidding for design and build work is intrinsically more expensive than for architect designed projects because of the costs of design work and the preparation of bills of quantities. However, Hughes' research suggests that traditional competitive tendering is actually up to four times more expensive than design and build. This is probably due to the higher levels of competition and lower tender success rates in traditional tendering.

Hughes indicates that tendering success rates are normally around 1 in 5 within larger contracting organisations but this will vary according to the type of work tendered for, the balance of competitive and negotiated work, the nature of the procurement arrangements for projects and the market circumstances at the time of tendering.

The average time spent on a major enquiry was found to be in the order of 400–500 man-hours per bid. Taking an average 5-week tendering period, this would suggest that a contractor employs between two and three people full-time on each tender.

Contractors' head office overheads vary of course but can be as low as 3–5% in a large company and thus tendering costs represent a considerable proportion of the total. However, without this overhead expense the contractor cannot win any work.

5.9 Pre-tender arrangements

Where the contractor has decided to submit a bona fide tender, the estimating, buying and construction team will need to be motivated into action.

In small and medium-sized companies, the senior personnel involved in the tendering process will normally be the managing director, the director responsible for estimating and surveying and the contracts manager. The main responsibility for putting the tender together will lie with the estimator/surveyor in conjunction with the contracts manager. A commercial/office manager will normally be responsible for the distribution of tender documentation from the client.

A typical organisation structure for such a business is shown in Figure 5.1 where it is assumed that the company has no separate buying or project planning facilities.

In medium and large companies, a functional organisation structure is common with separate departments for the estimating/surveying function, contracts and planning function and the administration of the business, as shown in Figure 5.2. In this type of organisation structure, the personnel involved in the estimating and tendering process will be:

- Senior estimator/buyer/estimator
- Contracts manager
- Tender planning engineer
- Commercial manager/office manager and clerical staff

Advice on contractual matters and commercial risk issues will also be sought from the company's quantity surveying department.

A large company may employ up to five or six estimators under the direction of a chief estimator. Standard procedures and estimating pro-formas are often developed based on computer spreadsheets. Information flow is controlled by the chief estimator or commercial manager.

An important aspect of the chief estimator's job is the management of the tendering process, which includes considering the workload of the estimating team. Figure 5.5 shows a simple bar chart which might be used to indicate the workload and key estimating dates over a monthly period.

Tender submission deadlines are invariably tight and the tendering process needs careful management because mistakes are likely to be costly. If procedures are not followed, the contractor might miss the submission date or lose the contract on a technicality. Worse still, the contractor might lose money should the tender be successful. Most competent contractors will have a system of pre-tender procedures and tasks will be delegated by the chief estimator to make sure things happen as they should. These tasks will include:

- Receipt of tender documents
- Registering the tender and allocating a tender reference number
- Circulating the tender enquiry summary
- Setting up the decision to tender meeting (tender launch meeting)
- Distributing documents to the tender team
- Arranging review meetings during tender
- Arranging adjudication meeting and submission of final bid

The chief estimator will also delegate responsibility for the preparation of the estimate and tender summary. This will be coordinated by the estimator but several other staff will also be involved. Their input will be as shown in Table 5.2.

5.10 The site visit report

Before starting to price the tender, the contractor should visit the site in order to satisfy himself of the prevailing conditions. The visit will usually be made by the estimator and perhaps a colleague such as the contracts manager.

As well as being a sensible precaution, the site visit is usually a condition of tendering and there is normally a 'sufficiency of tender' provision in the contract preliminaries. This means that the contractor is deemed to have satisfied himself, as far as he is able, with regard to the condition of the site and that the tender takes this into account. The following factors will directly influence the contractor's tender price:

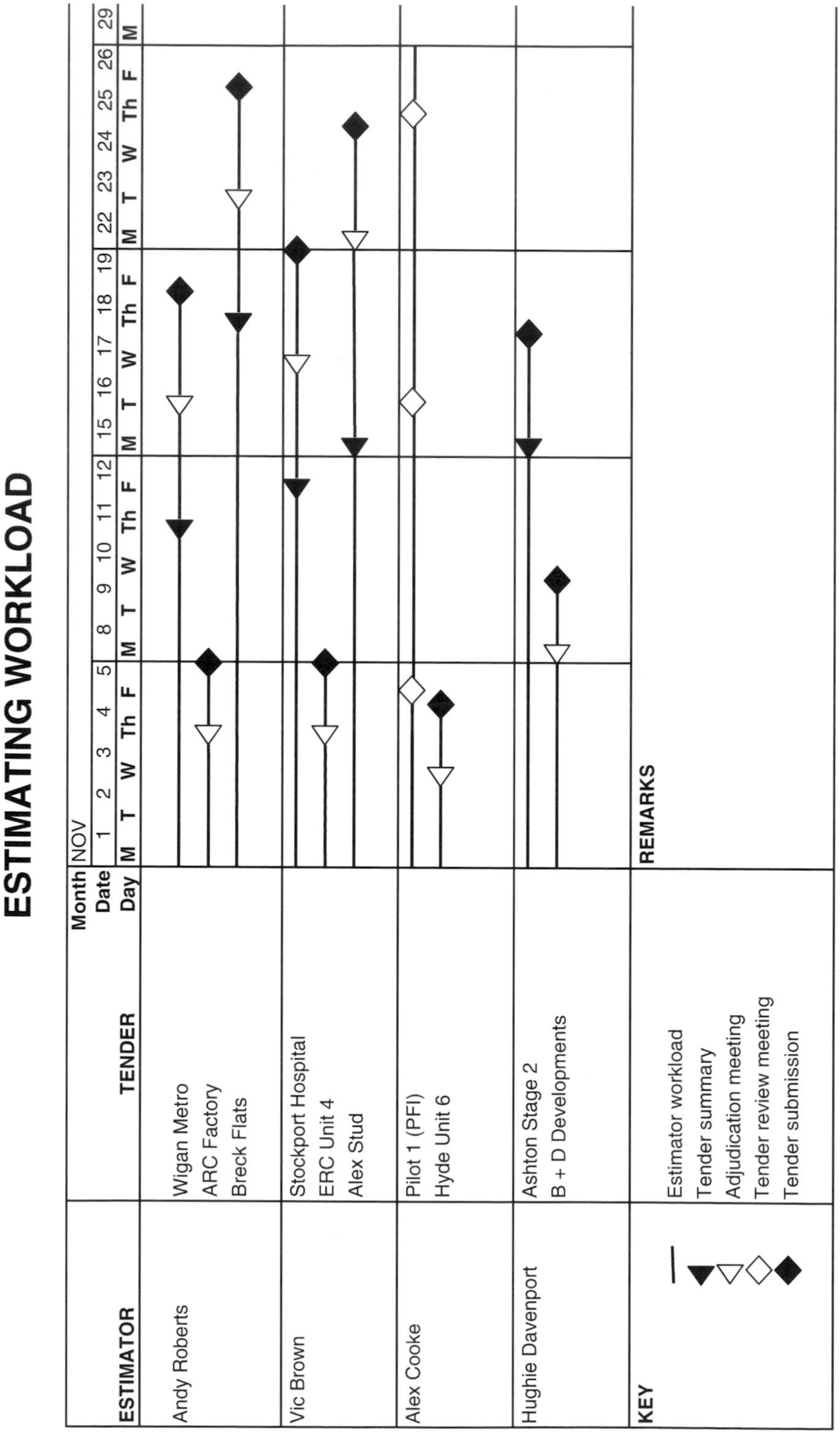

ESTIMATING WORKLOAD
ESTIMATOR
TENDER
Month NOV
Date 1 2 3 4 5 8 9 10 11 12 15 16 17 18 19 22 23 24 25 26 29
Day M T W Th F M T W Th F M T W Th F M T W Th F M
Andy Roberts
Wigan Metro
ARC Factory
Breck Flats
Vic Brown
Stockport Hospital
ERC Unit 4
Alex Stud
Alex Cooke
Pilot 1 (PFI)
Hyde Unit 6
Hughie Davenport
Ashton Stage 2
B + D Developments
KEY
Estimator workload
Tender summary
Adjudication meeting
Tender review meeting
Tender submission
REMARKS

Figure 5.5

Table 5.2 Estimating responsibilities.

Personnel	Responsibility
Estimator	Dealing with the management and pricing of the bills of quantities Dealing with subcontract and material price enquiries when there is no company buyer available
Contracts manager	Arranging to visit the site with the estimator Preparing the site visit report Preparing the method statement and discussing it with the estimator Preparing an assessment of the project plant requirements Advising the estimator on the requirements of the contract preliminaries Preparing an assessment of the pre-tender programme, in conjunction with the company planning engineer Preparing an assessment of the project safety requirements in response to health and safety legislation Arranging to view the project drawings, if not included with the tender documentation
Company planning engineer	Preparation of an assessment of the pre-tender programme Assistance in the assessment of the construction method statement
Senior management	Overseeing all stages of the estimate Liaison with the contracts manager with regard to major decisions Chairing the tender adjudication meeting and taking all necessary decisions on mark-up and profit additions

- Access to the works – effect on construction methods
- Access restrictions affecting the utilisation of plant
- Site topography, ground conditions and groundwater levels
- Distance to local tips for the disposal of material
- Provision of site security
- Restrictions imposed by adjacent buildings and services

Many of the large construction firms make use of a standard site visit report pro-forma, which acts as a check-list during the site visit and ensures that essential data is not overlooked. A model site visit report is provided in the CIOB Code of Estimating Practice (CIOB 1997).

Check-list

Points to be noted when preparing a site visit report include the following.

Names of parties involved in the contract

- Particulars of local authorities and planning and building control, etc.
- Details of statutory undertakings (water, electricity, telephones)

Site access

- Low bridges affecting the movement of plant and equipment to site
- Busy roads and potential offloading problems
- Traffic management on and around the site
- Specific site requirements for temporary works including access roads

Nature of site

- Topography – flat/sloping/extent of bushes and trees, etc.
- Site location relative to adjacent roads
- Other contracts in the immediate area
- Existing services locations

Existing buildings

- Extent of buildings to be demolished – site photographs prove useful when pricing
- Consider recording the condition of buildings on a digital camera
- Location of and dangers from buildings adjacent to the works – this may influence the cost of temporary works and excavation methods
- Details of any fly-tipped materials on site

General ground conditions

- Site surface conditions – adjacent excavations may provide evidence of ground conditions and water table levels
- Details of any watercourses crossing the site should be noted

Disposal of excavated material and waste

- Distance to local tips
- Tipping charges and landfill taxes
- Types of materials to be removed from the site – this will influence disposal costs, particularly if 'special waste' is involved
- Local skip-hire firms

Existing services

- Evidence of water and electricity supplies, drains and manholes
- Presence of telegraph poles and overhead cables

Site security

- Possible requirements for hoarding and site compounds
- The need for the provision of secure containers especially on inner city sites
- Likely requirements for security patrols and dogs

Local labour

- Assessment of the labour availability in the area
- Availability and quality of local subcontractors
- Telephone number and address of local Job Centre

Material suppliers

- Names of local builders' merchants and suppliers
- Local plant hire firms and size of plant fleet

Working space

- Space available for cranes and hard standings
- Site boundary constraints requiring crane 'easements'
- Lay-down areas for material storage
- Site parking for staff and operatives
- Vacant land nearby for rent if required

As a note of caution, many a contractor has lost a project because of the inability to carefully assess risk. Tenders are often won on the basis of items that the contractor does not include in the price, either missing them altogether or simply taking a 'flier'. Site ground conditions are particularly important and the effect on pricing of temporary earth support needs careful consideration. This particular item may be treated as an item of contractual risk within the tender.

The site visit provides vital clues to the presence of these risk factors and the estimator needs to pick them up and make proper provision in the pricing. Whether the directors decide to put them down to 'contractor risk' is another matter!

5.11 Tender enquiries to subcontractors and suppliers

Subcontractors

One of the key decisions facing the estimator is how to 'package' the work when sending out enquiries to subcontractors at tender stage. The problem is that the bills of quantities will normally have been prepared using one of the standard methods of measurement used in the industry, but the bills are not necessarily arranged conveniently for subcontract packaging (or 'parcelling' as it is sometimes called).

If work is packaged by following the bills of quantities work sections, enquiries may contain work activities which are outside the scope of the work normally carried out by a particular subcontractor. Conversely, where there are interfaces between packages, e.g. demolition and excavation, or groundworks and reinforced concrete work, there may be uncertainty as to what each of the subcontractors have priced into their tenders. Such interfaces can also result in disputes where none of the package contractors has included for a particular item of work or, alternatively, duplication may occur where several firms are responsible for the same item. Mistakes can easily be made here and the estimator must ensure that all bill items are priced one way or the other.

It is usually best to parcel work using normal trade demarcations of the industry, thus ensuring that each subcontractor will receive an enquiry for work with which he is familiar. Alternatively, the main contractor will send enquiries to firms which specialise in, say, formwork, rebar and concreting or excavation and drainage, by selecting appropriate parts of the bills of quantities to send to them.

An important consideration here is the need to obviate further subletting of the package to sub-subcontractors or even sub-sub-subcontractors which can lead to multi-level interface problems resulting in coordination difficulties and uncertainty as to responsibilities and liabilities on site. If well thought out and planned, packages can be let so as to avoid this situation thereby reducing the potential for confusion, mistakes and accidents.

Partnerships with preferred subcontractors are now common in the industry and Chapter 7 deals specifically with procedures for managing the supply chain at the tender stage and during the contract on both traditional and design and build projects.

Suppliers

Part of the estimating process concerns sending enquiries to material suppliers so that the estimator can price the measured items in the bills of quantities. Main contractors tend to use subcontractors for most site operations nowadays, but for certain trades the contractor will engage his own labour and price the work himself.

At tender stage, quantities for the main materials have to be extracted from the bills of quantities and this is done either by the estimator or by an estimating clerk. This is not straightforward as the bills of quantities contain items of finished work measured in m, m^2 or m^3, but materials are often supplied in different units. For instance, brickwork is measured in m^2 but supplied in packs and priced per 1000 bricks. The CIOB Code of Estimating Practice (CIOB 1997) indicates the information to be included in enquiries to material suppliers.

When the estimator receives prices from the various suppliers these have to be compared to see which is the most appropriate price to use. A comparison spreadsheet is usually used for this. The CIOB Code provides useful schedules for guidance.

5.12 Build-up of the estimate

Pricing is the sole responsibility of the estimator although on major projects there may be an estimating team involving several estimators, planners and contracts staff. Assistance may also be available from the company buyer, if the company possesses such expertise. Within the medium-sized organisation it is usual for the estimator to be responsible for sending enquiries to subcontractors and suppliers. The various sources of data available to the estimator are summarised in Figure 5.6.

The build-up of the estimate will involve the estimator in the following processes.

Build-up 'all-in rates' for labour

The 'all-in rate' for craft operatives and labourers varies from contract to contract due to varying conditions on site and the degree of supervision allowed in the build-up. The CIOB Code of Estimating Practice (1997) illustrates a typical build-up.

Pricing the bill sections

Bill sections may be built up in analytical form with a separate labour, material and plant inclusion for each bill item. Rates are usually established net with the percentage for overheads and profit added at the adjudication stage. Alternatively the estimator may use his own method of building up rates, based on his experience and knowledge of labour-only rates currently applicable.

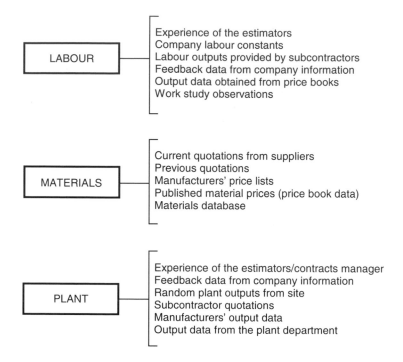

SOURCES OF ESTIMATING DATA

Figure 5.6

Analysis of the quotations

This is a comparative analysis of prices received from various sub-contractors and materials suppliers. Adjustments need to be made for differences in the quotations received to enable like-for-like comparisons to be made. For instance, allowances need to be made for items not priced, qualifications to the price, discounts allowed and facilities required from the main contractor for offloading, hardstandings, access, etc. (called attendances).

Build-up of the contract preliminaries

Preliminaries are the contractor's general cost items which do not form part of the permanent works. They include such items as site supervision, provision of temporary accommodation, on-site services and power, temporary access, hoardings, etc. Preliminaries items generally fall into two main categories: fixed costs (one-off charges) and time-related costs (those that vary with time). The estimator will often receive advice from the contracts manager and senior management when building up the preliminaries items.

Preparation of tender summary

This involves collating the estimate, analysing the overall price into labour, materials, plant and subcontractor elements and identifying other major items of cost such as preliminaries, prime cost and provisional sums, insurances and bonds. A tender summary sheet will be prepared by the estimator.

Attending the tender adjudication meeting

The estimator will present a report to management summarising the approach taken to the pricing of the tender and identifying the key risk issues and allowances which will need to be considered by the adjudication panel.

The estimator is usually the one who gets the blame for winning tenders that ultimately finish up losing money. It is so easy for site management to blame their own inefficiencies on bad estimating. The estimator always gets the negative feedback but is never told when his rates make money. Estimating is a no-win situation!

5.13 Pre-tender method statement

The pre-tender method statement will outline the sequence and method of construction upon which the estimate is to be based. It should indicate how the major elements of work will be dealt with, and highlight areas where new or alternative methods are being considered.

The method statement may be supported by details of gang sizes and outputs, plant and supervision requirements. While there are no strict rules to follow, for most projects the pre-tender method statement will be a simple outline document conveying the contractor's general approach to the construction methodology but with little detail. The tendering period is often too short to afford the luxury of spending much time on this stage of the process.

For civil engineering tenders, the converse may be the case as the items in the bills of quantities are often priced on the basis of operational pricing rather than the unit rate method of estimating. Greater reliance would therefore be placed on producing a well-developed construction method at tender stage as an aid to accurate pricing of both the permanent and temporary works.

Pre-tender method statements are particularly useful when the estimator or planner needs to assess the costs of any alternative proposals. It is a good idea to involve members of the contracts staff in the production of the pre-tender method statement to allow those ultimately responsible for undertaking the work to provide some practical input and ideas at the estimate stage, thereby developing a sense of ownership.

The pre-tender method statement is usually prepared by the contracts manager in the medium-sized contracting business but within the larger company input will also be available from the planning section.

Example pre-tender method statement

A subcontractor has received a tender enquiry to price the concreting operations for a six-floor Victorian factory building. The tender documents require the concrete to be laid 100 mm thick and each floor is 500 m² in area. Access to each floor is to be via a central goods lift. Alternative proposals for placing the concrete may be submitted with the tender.

The contracts manager for the subcontract firm will prepare the method statement in order to advise the estimator of proposed methods for inclusion in the estimate. The following points should be considered:

- The main operations to be included in the method statement
- Availability of resources – labour gang size, plant and supervision
- Approximate quantities
- Output rates per gang per day
- Bay or pour sizes
- Any reasoned alternatives

Developing the pre-tender method statement – written format

Operation   Handle and place concrete in floor bays

Option 1 – Place concrete using dumpers

Method
Ready mixed concrete supply to rear entrance of factory at ground floor level. Discharge concrete into front discharge dumpers and transport to appropriate floor level via central goods lift.

Discharge concrete into floor bays and place by hand. Tamp floor slab with timber tamping board and float off with patent steel levelling screeder to obtain required tolerances.

Concrete to be placed in 5 m wide bays running full length of building. Alternate bay construction to be used. Floor surface to be covered with polythene to aid curing for a 1-day period.

Two bays per day to be poured .

Resources
Plant
3 No. 1 cubic metre capacity dumpers
1 No. 5 metre timber screed board
1 No. steel levelling screed
1 No. petrol vibrator

Labour
3 No. dumper drivers
3 No. labourers placing concrete
2 No. labourers levelling and finishing
1 No. foreman supervisor

Output and quantities
Quantity – 50 m^3 class C concrete per floor
Quantity poured per day – 25 m^3
One floor to be placed in 2 days

Option 2 – Place concrete using mobile concrete pump

Method
Mobile concrete pump to be located in rear access yard. Direct discharge of ready mixed concrete into pump. Concrete to be pumped through existing window openings to floor bay location. Place concrete and tamp into position. Additional cost of using 'flowable' concrete to be considered. Concrete to be placed in one continuous bay per floor using temporary screeds between bays.

Level off each bay using vibrating screed board and steel levelling screed. Protect as before.

Resources
Plant
1 No. hired concrete pump
1 No. vibrating screed board
1 No. steel levelling screed
1 No. petrol vibrator

Labour
1 No. labourer with concrete pump
4 No. labourers placing concrete
2 No. labourers levelling and finishing
1 No. foreman supervisor

Output and quantities
Quantity 50 m^3 per floor
One floor is to be poured per day

FACTORY REFURBISHMENT — TENDER METHOD STATEMENT

OP. No.	OPERATION	QUANTITY	METHOD	RESOURCES		NOTES
				PLANT	LABOUR	
1	Concreting to factory floors (floors 1–6)	50 m³/floor	**OPTION 1** Place concrete using front discharge dumper trucks. Goods lift to be used as access from ground floor level. Place concrete in 5 m wide bays and tamp level. Steel levelling screed finish.	3-Dumpers 1–5 m Tamp 1-Steel screed 1-Vibrator RMC Supply	3-Drivers 5-Labourers 1-Supervisor	Output per day 25 m³
2	Concreting to factory floor	50 m³/floor	**OPTION 2** Place concrete using mobile concrete pump located in rear yard area. Pump concrete through window openings at each floor level Continuous bay pour – 1 pour per floor.	1-Vibrating screed 1-Steel screed 1-Vibrator RMC Supply	6-Labourers 1-Labourer at pump 1-Supervisor	Output per day 50 m³

Figure 5.7

Figure 5.7 illustrates an alternative, tabular presentation of the pre-tender method statement. This has the advantage of showing the operational resource requirements in a somewhat clearer format.

The estimator or planning engineer would be responsible for assessing the costs of the alternative proposals and reporting his findings to the contracts manager. From a planning and speed of construction point of view, Option Two will evidently prove to be the more desirable method for inclusion in the tender. The technique of using operational estimating techniques based on method statements is explained by Smith (1995) who illustrates an example build-up of alternative method proposals. Operational estimating techniques are also covered in the CIOB Code of Estimating Practice (1997).

5.14 Preparation of pre-tender programme

In the majority of construction contracts it is usual to have an agreed period within which the works are to be completed. This gives certainty to the contract and allows key events to be triggered. These include extensions of time for completion and the charging of liquidated and ascertained damages for delay.

The period for completion will either be stipulated within the tender documents or stated in the contractor's tender offer. On occasion there will be a stipulated period and also an opportunity for tendering contractors to state the period within which they propose to carry out the works as an alternative for consideration by the client.

As part of the tendering process, the contractor will have to think about how the job will be built so that the tender figure is realistic and technically achievable. The contractor cannot simply rely on the contract period stated in the tender documents for building up the costs of preliminaries and major items of plant.

A pre-tender programme is essential in order that the contractor may be reasonably sure that he will be able to complete the project on time, as there could be severe financial consequences for failure to do so. For instance, time overruns can result in additional costs and there may also be liquidated and ascertained damages to pay to the employer for failure to hand over the building on time.

The pre-tender programme will be presented in sufficient detail to enable the estimator to price the time-related elements of the tender and enable the adjudication panel to judge the time- and method-related risks attached to the project.

When the programme has been thought through, it may be possible to tender on a reduced time-scale to that which is stipulated in the tender documents. This will facilitate a reduction in the time-related elements of the preliminaries and thus give the contractor a competitive edge over other tenderers.

On the other hand, the contractor may find that the stipulated period is insufficient to complete the job safely and to the desired quality. It is no use tendering for a contract on the basis of a 30-week programme if the contractor considers that he needs 38 weeks to construct the project. In such circumstances, the contractor may qualify his tender to this effect. This is done by enclosing a letter of qualification with the tender submission.

However, it might be a condition of tendering that qualifications are not allowed. In this event, the contractor will have to make an allowance in his bid for the cost of overrunning the programme (i.e. extra preliminaries) and he would also be wise to make an allowance for any liquidated and ascertained damages that might be charged to him for failure to complete on time.

Where the contract period is specified, it is necessary for the contractor to assess whether this period is realistic or not.

At tender stage the pre-tender programme is normally presented in bar chart format. The programme will cover the major stages of the work to be undertaken based on the information available from the tender drawings. The utilisation of linked bar chart techniques based on project management software will enhance the programme presentation at this stage. It is now becoming common practice for the pre-tender programme and method statements to be submitted with the priced bid.

Where a project is being negotiated, it may be the practice of the contractor to present the pre-tender programme in network or precedence format. This however depends on the complexity of the project and information available.

Figure 5.8 shows a typical pre-tender programme for a small factory project which would normally be prepared in bar chart format using project management software.

5.15 Build-up of contract preliminaries

The purpose of the contract preliminaries is to include for the pricing of items which cannot be reasonably included in the measured work. Contract preliminaries are priced on a fixed and time-related basis. Certain preliminary items relate to a one-off or fixed charge to the project, i.e. the erection of cabins and site accommodation, while other items are directly related to the contract duration, i.e. the utilisation and weekly hire costs of using a tower crane.

The division of the preliminaries costs into fixed and time-related elements is directly reflected in additional charges to the client in relation to extensions of time and claims for loss and expense. It has been shown that tender success rates improve where management involvement is evident in the build-up of the contract preliminaries.

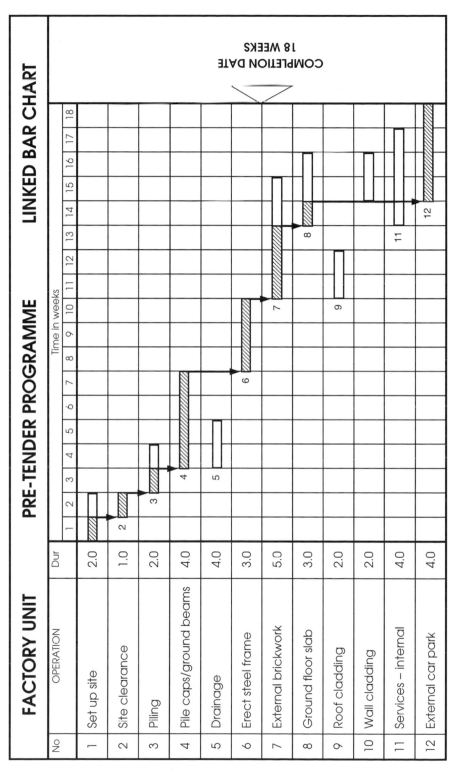

Figure 5.8

The CIOB Code of Estimating Practice (1997) includes an extensive project overheads schedule to assist in the build-up of preliminaries costs. This ensures a logical approach to the assessment of project overheads.

Preliminaries check-list

Employer's requirements

- Time-related costs include allowances for site accommodation and telephone facilities for the use of the architect and other consultants; also, hire costs of a computer facility if specified in the documents
- Fixed costs include the erection and dismantling of accommodation, fitting out and decorating, furniture and fittings, technical and surveying equipment, telephone installation and site notice boards

Contractor's management and site staffing

- Time-related costs include site staff allocated to the project, including site manager, trades supervision, site engineering staff and site support staff. In certain organisations it may be policy to treat staff servicing the site as part of the head office overheads – contracts managers, surveying and planning personnel, etc. This depends on the policy within the organisation and their approach to the pricing of project overheads

Site accommodation costs

- Time-related costs include hire charges for all required site accommodation, including the servicing of such facilities
- Fixed costs include the delivery to site, erection, fitting out and the dismantling and removal on completion

Attendance on domestic and nominated subcontractors

- Time-related costs include unloading and distribution of materials, attendance on domestic subcontractors and progressive clearing of building waste
- Fixed costs include the cleaning out of the building on completion of the works

Facilities and services associated with the works

- Time-related costs include weekly costs of telephones, office equipment and stationery, safety, health and welfare provision, power

and lighting for offices and site security costs per week. This may also include such items as the removal of site rubbish, drying out buildings and office cleaning services
- Fixed costs include installation and connection charges relative to the above. The majority of the costs will be involved in temporary service connections

Temporary works costs

- Time-related costs include maintaining, cleaning and servicing access roads, hard standings, compounds and hoardings. Moneys may also be included for complying with traffic regulations and work in connection with attendance on pumping and de-watering
- Fixed costs include the initial installation and removal of items such as access roads, hard standings, compounds and hoardings. Items in relation to pumping and de-watering may be included separately

Mechanical plant

- The selection of plant for a project will have a direct effect on pre-liminaries costs and ultimately the competitiveness of the tender. The pre-tender method statement requires careful review in the light of proposed alternatives, and cost exercises may need to be undertaken. Installing a track for a tower crane may prove expensive and must not be considered lightly at estimate stage. Consideration must also be given to the provision of power supplies to large items of plant. All these factors may force the contractor into taking more risk at the tender stage
- Time-related costs include the hire charge per week of cranes, hoists, mixers, vibrators, compactors, dumpers, access equipment and mobile elevating working platforms (MEWPs), tele-handlers and fork-lift trucks and site vehicles for the collection of goods. Plant items such as excavators may be included in the built up rates. The contractor's approach to the inclusion of mobile concrete pumps is debatable, as this may be treated as a preliminary cost in the mechanical plant section or included in the unit rate for concrete work
- Fixed costs include the transport of the plant to the site, cost of receiving and installing and removal on completion. This should include any additional costs in providing temporary works such as the track or base for tower cranes

Non-mechanical plant (or small tools and equipment)

- This includes such plant items as bar bending machines, and provision of saw benches, ladders, wheelbarrows, road barriers and lamps, slings and chains, skips and surveying equipment. The main item included here is scaffolding, which will carry the bulk of the cost in this section

- Quotations will normally be obtained for scaffolding as part of the subcontract enquiry process. The contractor may also include as a fixed cost items such as power tools, including the transportation to and from site. Many contractors tend to price non-mechanical plant as a percentage addition on the labour cost (typically 1–2%)

The preliminaries are an important element of the contractor's tender and may represent between 5% and 15% of the tender sum. However, if the contractor included every conceivable item in the contract preliminaries section he would never win a competitive tender. Clearly, the contractor needs to be competitive and it is common for directors to shave off money from the preliminaries to help the contractor win the contract. This is often done in conjunction with considerations of the project programme, especially where the contractor decides to tender on a shorter duration than stipulated in the tender documents.

Contractual risk must however come into the equation somewhere. As previously stated, the contractor tends to win the tender on items he has not included rather than those he has.

5.16 Response to pre-tender health and safety plan

The Construction (Design and Management) Regulations 1994 require the planning supervisor to ensure that a health and safety plan is prepared and provided to the contractor before arrangements are made to carry out or manage construction work. For most projects this takes the form of a pre-tender health and safety plan which the successful tenderer then develops into a construction phase health and safety plan before starting work on site.

The pre-tender health and safety plan should arrive with all the other tender documents so that tenderers may consider its implications at the time the contract is priced. This enables the contractor to take into account particular site constraints and health and safety risks when building up his tender price, and to plan and programme the works accordingly. For instance, the contractor may have to allow in his pricing for dealing with design risks identified in the health and safety plan or there may be specific health and safety requirements or restrictions in the health and safety plan which may have a cost implication.

It is clearly unrealistic to expect contractors to fully develop their health and safety proposals at tender stage, both in terms of the effort needed and the time constraints of the tender period. However, asking for a response to the pre-tender health and safety plan to be submitted with tenders is a good way to confirm the competence of contractors in the health and safety management for a specific project, and whether sufficient resources for health and safety have been allocated in the tender. For instance, an abnormally low price may indicate lack of adequate health and safety provision in the tender, while an overoptimistic programme may mean that insufficient time has been allocated to the proper planning and implementation of health and safety management provision on the site.

This is not a requirement of the Construction (Design and Management) Regulations 1994, but it is becoming common good practice in the industry to ask for this response in the tender enquiry documents. The contractor's tender stage response should cover such issues as:

- Health and safety policy
- Evidence of adequate health and safety provision for the project
- Resources available to control and manage the major health and safety risks
- Evidence of competence to comply with health and safety legislation

Tender enquiries from contractors to potential sub-contractors should also include the pre-tender health and safety plan, or at least relevant extracts from it, so that any pricing implications in the health and safety plan can be included in subcontract quotations.

The contractor's tender stage response to the pre-tender health and safety plan will probably be insufficient to allow work to start on site, and therefore time needs to be allowed in the overall planning of the project between contract award and site possession to allow the contractor to develop the health and safety plan and satisfy the requirements of the CDM Regulations.

5.17 Management adjudication of the estimate

When the estimator has completed the pricing, a summary sheet will be prepared which will indicate a bottom-line figure. This is the total estimate for the project. Prior to submitting the tender at the appointed time, the directors will have to consider the estimator's calculations and make some decisions, including:

- Is the estimate accurate?
- Have the items been priced sensibly?
- Are the subcontract prices used achievable?
- Can the prices used be bettered if we win the contract?
- Is the pre-tender programme realistic?
- Can the programme be shortened?

- What are the risk issues related to this job?
- How much does the company need the work?
- What is the level of competition for the work?
- Does the tender documentation offer any commercial opportunities?

When these questions have been answered, the directors will be in a position to:

- Agree with the estimator's price
- Add money to the price, or
- Reduce the price

Whatever the decision, this will be the tender sum which will be offered to the employer via the form of tender. If the tender is accepted, the contractor will have to live with the decisions made and get on with building the project for the price offered. The pressure will then be on the site team to work within the budget and return the desired margin (profit).

The CIOB Code of Estimating Practice proposes a model form for the estimate summary, analysis and report.

Adjudication check-list

The following check-list provides an aide-memoire to assist the tender adjudication process and make sure that key issues are not overlooked. All information generated in the build-up of the tender figure needs to be included. For a decent-sized tender the amount of supporting documentation may be substantial, as follows.

General information

- Tender enquiry form including details of parties involved in the contract
- Site visit report
- The estimator's report
- The quantity surveyor's commercial risk assessment
- Knowledge of competitors
- Details of current and future contractual commitments

Pricing data

- Net cost analysis including the value of work to be sublet
- Analysis of subcontract and materials prices
- Schedule of prime cost and provisional sums
- Schedule of project overheads
- Tender summary

Pre-tender planning data

- Overview of construction methods allowed in estimate
- Planning report from the planner/contracts manager comprising comments on the contract period given by the client and the feasibility of the pre-tender programme
- Pre-tender programme (linked bar chart)
- Pre-tender method statements

Risk issues

- Ground conditions report
- Temporary works allowances
- Design liability (if any)
- Summary of contract conditions including any amendments to the standard form
- Details of the damages for late completion
- Details of insurance requirements and bonds
- Working capital implications for the contractor's borrowings
- Any cash funding requirements required

Finally, the contractor must consider the desired mark-up in the context of the quality of tender information, the technical and contractual risks identified in the tendering process and the construction time and methods allowed. Profit percentages could be anything from 1% to 20% depending on how keen the contractor is to win the contract.

In a very keen market, the contractor may look for ways of taking money out of the tender by anticipating discounts and buying savings from subcontractors and suppliers or by gambling on possible claims and other commercial opportunities that might arise during the contract. At this stage it may be better to hire a crystal ball!

A percentage addition is normally applied to cover head office overheads. This might be in the range of 5–10% depending on the size of the organisation.

A typical agenda for the adjudication meeting is given below:

AGENDA FOR TENDER ADJUDICATION MEETING

Project title
Tender no.
Date

Personnel present
Project details
- Client
- Description of works
- Budget (if known)

- Form of contract
- Liquidated and ascertained damages
- Completion period

Project strategy
- Construction methods
- Programme
- Site establishment
- Technical issues
- Health, safety and environmental issues

Tender strategy
- Market conditions
- Risk issues

Tender analysis
- Net cost of labour, plant and materials
- Subcontractors and discounts
- Preliminaries
- Temporary works
- Mark-up

Risk management
- Risk register
- Payment terms
- Insurances
- Onerous contract conditions
- Programme

Submission documents
- Priced bills of quantities
- Contract sum analysis
- Response to pre-tender health and safety plan
- Pre-tender programme

Tender sign-off
- Risk allowances
- Final tender adjustments
- Final tender figure agreed

Once the final adjudication decisions have been made, the tender summary sheet will have to be revised and any changes expressed in the tender submission. This may be a fully-priced bill of quantities or simply a figure stated on a form of tender.

The form of tender will have to be signed by a responsible person and submitted in the appropriate envelope. If alternative proposals have been invited from the contractor, these should be clearly stated.

Quite frequently contractors are asked to submit two envelopes containing both a price and a quality bid. This enables the client team to judge the contractor's approach to the management of the project as well as the price. The quality bid is usually opened first and judgements are made on pre-selected criteria to give a quality score for the tender. If the quality bid is satisfactory then the price envelope is opened. Choice of the successful tenderer will then be made on the basis of a weighting of both price and quality (say 60/40).

5.18 Analysis of tender performance

The application of bidding theory is not common in construction but it is not altogether unheard of either. This is the practice of judging tender performance in a statistical and analytical form. However, many practitioners consider the technique to be of little benefit and purely the preserve of academics. Bidding theory is only relevant in stable market conditions, where the contractor is bidding against known competition in the market place. In the current volatile construction climate, this is rarely the norm. Changes in procurement procedures have led contractors to enter into partnering arrangements, with more emphasis on negotiated contracts, and hence emphasis on bidding strategies is less relevant.

Once a bid has been submitted and the contractor has committed himself to a contract price, it is no good reviewing what he should have done in the light of other competitor bid prices. 'If only we had done this or adjusted that' is crying over spilt milk and the only alternatives are to sack the estimator – because surely it is now all his fault! – or just get on with the job. Whatever the case, management rarely takes the blame for adjudication decisions.

On acceptance of the contractor's bid price, procedures will be implemented to check the priced bill, agree any bills of reduction if tenders are over the client's budget, and arrange for contracts to be signed. Letters of intent may be issued so that the contractor can prepare designs or order materials with long lead-times.

Before commencing work on site, the pre-contract planning process must be started and this is dealt with in Chapter 6.

References

Bentley, J. I. W. (1987) *Construction Tendering and Estimating*. E. & F. N. Spon.

Brook, M. (1993) *Estimating and Tendering for Construction Work*. Butterworth-Heinemann.

CIOB (1997) *Code of Estimating Practice*, 6th edn. Chartered Institute Of Building.

CIOB (undated) Construction Papers Nos. 7, 11, 15, 39. Chartered Institute of Building.

Hughes, W. (2004) Are bidding costs wasted? *Construction Manager*, Journal of the Chartered Institute of Building, January.

Smith, A. J. (1995) *Estimating, Tendering and Bidding for Construction*. Macmillan.

6 Contract budgets and cash flow

6.1 Contract budgets

It is an essential function of management to prepare forecasts in order to establish a plan for the future of the business. Without a plan against which to monitor performance, management has no control and no business can be run successfully without keeping close control over the day-to-day finances. This involves providing the information necessary for keeping managers informed so that decisions can be made about how the business should react to current circumstances.

Businesses are living entities (going concerns) and therefore money is coming in and going out all the time. There are wages and suppliers' and subcontractors' bills to pay and moneys are being received for contract work. This activity cannot be allowed to happen willy-nilly because the business will soon be in a mess. Contractors, therefore, have to prepare forecasts and these are required for many aspects of their business as well as the contract work undertaken. For instance, the company has to plan for the next year's trading in order to make sure that sufficient work is obtained to enable the company to meet its commitments and keep going. This is very much tied in to the estimating, planning and quantity surveying functions within a construction company.

6.2 Definitions

Financial plans are called budgets and the process of making, monitoring and adjusting them is called budgetary control.

Hall (1974) identifies budgetary control as an important management technique used for the purpose of controlling income and expenditure. Control is achieved by preparing budgets relating to the various activities of the business, and these provide a basis for comparison with actual performance.

Alternatively, Harper (1976) defines a budget as a monetary cost plan relating to a period of time, while Pilcher (1975) indicates that a budget is a financial plan for the contract as a whole and is a financial version of the programme.

The budget may be expressed in tabular or graphical format such as S curves, line graphs or histograms, although in practice managers tend to favour graphs as a way of expressing data.

6.3 Types of budgets

Harris and McCaffer (2000) identify the various types of budgets applicable to contracting organisations. These include:

- **Operating budget** – staffing, overheads, labour, materials, subcontractors
- **Annual sales budget** – anticipated turnover including existing contracts
- **Capital expenditure budget** – spending on plant and equipment, development land, etc.
- **Cash flow budget** – a forecast of the movement of moneys in and out of the business in order to determine company borrowings (working capital requirements)
- **Master budget** – a forecast balance sheet and profit and loss account

The budgets produced at head office level will be based on budgets relating to specific contracts or developments, built up to show the picture for the company overall. Developers or contractor/developers need to anticipate future borrowings very carefully as the working capital requirements for schemes relying on sales of completed units are particularly heavy.

A housing developer will prepare an annual sales budget in order to monitor actual sales against a predetermined plan, but a general building contractor may be more concerned about forecasting his cash flow. In this case, cash flow budgets would directly help with cash control and also identify working capital requirements. Budgets may also relate to turnover forecasts or the value of cumulative tender enquiries received. Capital budgets lay down the planned requirements for the long-term survival of the business.

At project level, contractors need to forecast the amount of work they expect to carry out each month (the value) and what the expenditure on wages, materials, plant, etc. is likely to be (the cost). For forecasting purposes, the difference between value and cost is margin (overheads and profit).

This process facilitates control over payments made by the client under the contract and enables the contractor to assess how the contract is performing financially. These budgets will relate to the contractor's tender using the priced bills of quantities and taking account of any tender adjudication adjustments.

6.4 Preparing budgets

The procedures involved in the preparation of a budget are outlined by Harper (1976) who suggests five stages of development:

(1) Prepare the forecast
(2) Consider the company policy
(3) Compute the budget by expressing resources, quantities or values in monetary terms to form the initial budget
(4) Review the forecast, policies and the initial budget until an acceptable budget emerges
(5) Accept the budget at all levels of management. Ensure that it has the company's backing from the top down

The budget is now established.

Figure 6.1 indicates the above stages diagrammatically, together with the principles of budgetary control applied to the control cycle for a sales, cumulative value forecast or cash flow budget.

6.5 Control procedures

Control procedures must now be established in order to monitor the budget at clearly defined time intervals. These may be daily, weekly, monthly, quarterly or annually. Management will then be able to take corrective action where variances from the budget occur. Management must act where variances are apparent at each review date, whether they are positive or negative. Any director can shout when it is obvious that things have gone wrong, but the astute manager is the one who analyses variance trends and applies decisive action at the appropriate time.

The application of variance analysis forms part of the budgetary control process. Examples of the analysis of contract variances are given in Chapter 9.

Variance analysis is the matching of actual performance with the forecast in order to assess the difference, whether it be positive or negative. It is the responsibility of management to highlight variances, report on the findings, and implement action the effect of which may often prove painful to subordinates. Failure to meet targets may have a damaging effect on morale within the organisation and questions will be asked about the initial forecasting approach and methodology. An easy way out at this stage is for management simply to admit that the forecast was wrong and to discharge the person responsible for preparing it. This is the usual escape route and gives everyone the impression that decisive action has been taken.

BUDGETARY CONTROL PRINCIPLES

STAGES IN PREPARING A BUDGET

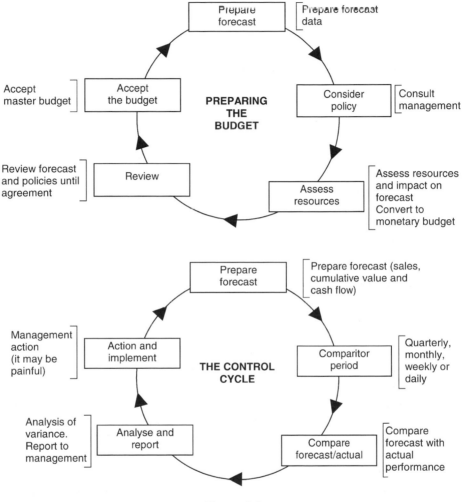

Figure 6.1

For budgetary control to be an effective tool in the management of the business, forecasting must be realistic and achievable.

6.6 Forecasting contract value

When assessing the financial requirements of a project, or a number of projects to be undertaken over an annual period, the contractor must ask the following questions:

- What is the basis of the forecast?
- What is the basic information required?
- How does timing affect the monetary requirements?
- How accurate does the forecast need to be?
- How does the contractor monitor the cash requirements during the project?
- Will the budget information help to convince lenders that the cash funding requirements have been realistically thought through? A slick presentation at this stage may help, but this is doubtful

Various techniques are available to help but their usefulness depends on what stage has been reached and the information available. An overview of these methods of forecasting funding requirements will now be outlined.

S curves

Various types of S curves are available on which to base a cumulative value forecast. The S curve is best applied to projects where the provisional sums are evenly spread though the contract, as the technique does not allow for high expenditure provisional items early in the project period. The basis for this method of forecasting is the presumption (established by research) that the cumulative expenditure on any construction project normally approximates to an S-shaped curve. Consequently, if we take each monthly payment on a project and add it to the total of previous payments, a cumulative S-curve will be described.

The quarter–third approximation is a geometric curve which has been found to give a reasonable assessment of value, while other approximations based on the ogee curve give a similar range of cumulative values. The $^1/_4$:$^1/_3$ rule gives a good approximation of this curve. The basis of the rule is that:

- $^1/_4$ of total expenditure will be made during the first $^1/_3$ of the project time-scale
- a further $^1/_2$ of total expenditure will be made during the middle $^1/_3$ of the project
- the final $^1/_4$ of total expenditure will be made during the final $^1/_3$ of the project time-scale

It works by plotting a graph and Figure 6.2 shows the basic principles of the quarter–third approximation. Figure 6.3 gives a cumulative value forecast using this method for a contract lasting 6 months with a value of £160 000.

S CURVE APPROXIMATION
1/4–1/3 RULE

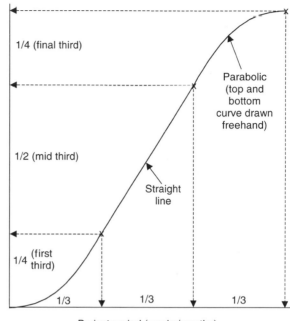

Project value (£000)

1/4 (final third)

Parabolic (top and bottom curve drawn freehand)

1/2 (mid third)

Straight line

1/4 (first third)

1/3   1/3   1/3

Project period (weeks/months)

PRINCIPLES

1/4 value expended in first 1/3 of project period

1/2 value expended in middle 1/3

1/4 value expended in final 1/3 of project period

Figure 6.2

Cumulative percentage value

This method involves preparing a forecast of the cumulative percentage value per month based on data analysed from similar types of project. The purpose is to produce a reasonable S curve approximation at the feasibility or tender stage of the project. This method is illustrated in Figure 6.4 which shows a tabular display of the cumulative percentage value forecasts for projects lasting up to 12 months.

The table in Figure 6.5 shows data abstracted from Figure 6.4 for a contract with a value of £160 000 and lasting 6 months. The appropriate contract duration is chosen along the horizontal scale and the estimated

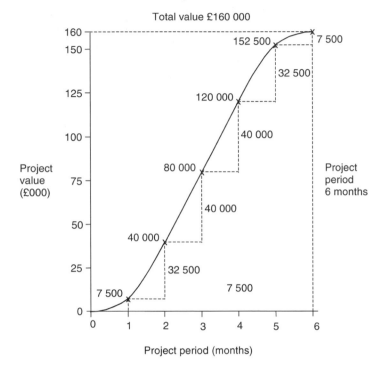

S CURVE PRESENTATION

Project value £160 000

Contract period 6 months

Total value £160 000

Project value (£000)

Project period 6 months

Project period (months)

VALUE FORECAST (Cumulative)			
Month	Cumulative value	Month	Cumulative value
1	7 500	4	120 000
2	40 000	5	152 500
3	80 000	6	160 000

Figure 6.3

percentage of the total contract value for that month of the project is read off vertically.

This approximation gives a slightly lower cumulative value forecast than the quarter–third method, as shown by the comparison of values and curve profiles in Figure 6.5. The tabular display in Figure 6.4 is ideal

PROJECT PERIOD (months)

Cumulative % value complete forecast

MONTH	1	2	3	4	5	6	7	8	9	10	11	12
0	0	0	0	0	0	0	0	0	0	0	0	0
1	100	45	24	18	12	9	7	6	5	5	4	3
2		100	76	45	30	24	18	14	12	11	9	7
3			100	79	60	45	35	27	23	20	17	13
4				100	82	67	55	45	37	31	27	22
5					100	85	72	63	53	45	38	33
6						100	88	77	67	59	51	45
7							100	90	80	72	64	56
8								100	92	83	75	66
9									100	92	85	76
10										100	93	85
11											100	93
12												100

HOW TO USE

Read off cumulative % value relative to contract duration, i.e. 6 month contract

CUMULATIVE VALUE

Month	Cumulative %
1	9
2	24
3	45
4	67
5	85
6	100

Figure 6.4

CUMULATIVE VALUE FORECAST

Contract value £160 000
Contract period 6 months

Cumulative value forecast (from % graph)

Month	Cumulative %	Cumulative value	Compared with 1/4–1/3
1	9	14 400	7 500
2	24	38 400	40 000
3	45	72 000	80 000
4	67	107 200	120 000
5	85	136 000	152 500
6	100	160 000	160 000

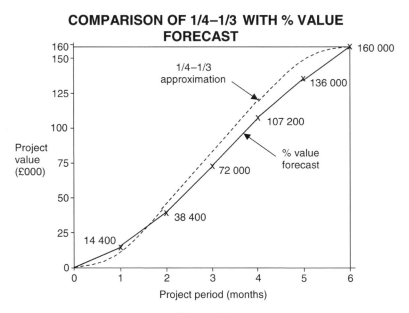

COMPARISON OF 1/4–1/3 WITH % VALUE FORECAST

Figure 6.5

for displaying as a spreadsheet, as this allows the analysis to be speedily undertaken and displayed in line graph form.

The idea of percentage value analysis relative to time may be developed by collecting empirical data for different types of projects – schools, housing, factories, office developments, etc. In this way a series of different S curve profiles may be developed for forecasting the cumulative project values at feasibility and tender stages.

Using the bar chart programme

This programme may be used to express the rate of value accumulation during the project. This may be achieved by analysing the contract estimate into an operational format in order to allocate the moneys to a linear time-scale on the bar chart.

As the project develops and further information becomes available, a more realistic cumulative value forecast may be prepared from the pre-tender programme or master programme. This enables the value forecast to be directly linked to the sequence of construction operations. The cumulative value forecast is therefore better prepared at this stage of the project as it provides a tool for controlling the project. At monthly intervals, the actual cumulative and forecast values may be matched as part of the company's monthly cost reporting procedures. The cumulative value forecast may also be used for assessing the client's and contractor's cash funding requirements. It is often a requirement of the tender submission that a cumulative value forecast is provided at the tender stage of the project in order to assist the client's assessment of his funding requirements.

Figure 6.6 illustrates a bar chart for a factory development project, showing the weekly and cumulative forecast values at each of the planned valuation dates. A cumulative value–time forecast based on the programme is shown in Figure 6.7. If the relationship between value and time is represented by a nearly straight line, as in this case, then so be it. The cumulative value forecast normally follows the basic S curve, except where high values are expended on such items as erecting steelwork early in the project period.

It must be pointed out at this stage that the cumulative value forecast is only as good as the accuracy of the planning forecast. If the planning is too ambitious, then so will be the resulting value forecast. It is the responsibility of management to overview the budget forecasts and ensure that they are attainable. All levels of management thrive on achieving success and meeting targets. The achievement of budget forecasts during a project helps establish a team approach to ensuring a successful project.

The good-guess method

This method is often used as a last resort when all else has failed to produce a satisfactory forecast. With practice and experience, preparing forecasts based on intuition may produce acceptable results. All too often though, 'well I guessed it' is the excuse given by the surveyor who is unable to substantiate his forecast when asked for the build-up.

FACTORY PROJECT – BAR CHART PROGRAMME – CUMULATIVE VALUE FORECAST

Time in weeks

OPERATION	BUDGET	1	2	3	4	5	6	7	8	9	10	11	12	13	14	15	16	17	18	19	20
Establish	5 000	5																			
Piling	10 000		10																		
Caps & g.beams	21 000			7	7	7															
Drainage	6 000		2	2	2																
Erect frame	20 000						10	10													
Roof cladding	14 000								7	7											
Ext. brickwork	30 000									6	6	6	6	6							
Floor slab	12 000													4	4	4					
Internal services	20 000															5	5	5	5		
External works	16 000																	4	4	4	4
Preliminaries	10 000	0.5	0.5	0.5	0.5	0.5	0.5	0.5	0.5	0.5	0.5	0.5	0.5	0.5	0.5	0.5	0.5	0.5	0.5	0.5	0.5
Weekly value (£000)		5.5	12.5	9.5	9.5	7.5	10.5	10.5	7.5	13.5	6.5	6.5	6.5	10.5	4.5	9.5	5.5	9.5	9.5	4.5	4.5
Cumulative weekly value		5.5	18.0	27.5	37.0	44.5	55.0	65.5	73.0	86.5	93.0	99.5	106.0	116.5	121.0	130.5	136.0	145.5	155.0	159.5	164.0
Monthly value					37.0				73.0				106.0				136.0				164.0
Valuation periods		1 →				2 →				3 →				4 →				5 →			

20-week project period

Project value £164 000

Figure 6.6

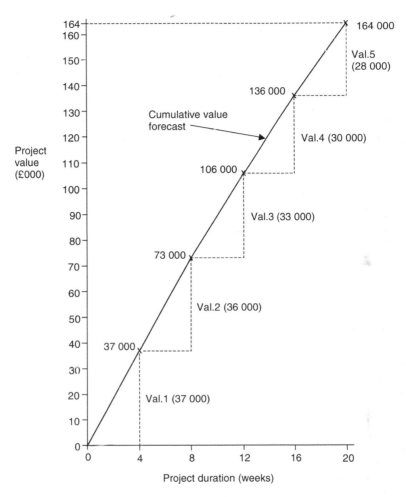

VALUE FORECAST FROM PROGRAMME

Figure 6.7

6.7 Credit terms in construction

Construction is largely a credit-based industry where customers expect a period of deferment before having to pay for goods or services received.

Traditionally the construction client expects the contractor to carry out a month's work which is then valued and certified and the certificate is later honoured by cheque. The average period of credit under a JCT 98 contract is illustrated in Figure 6.8 but this period can be much longer under other forms of contract. Contractors enjoy even better credit facilities from their subcontractors and suppliers.

The industry culture of late payment of subcontractors and suppliers

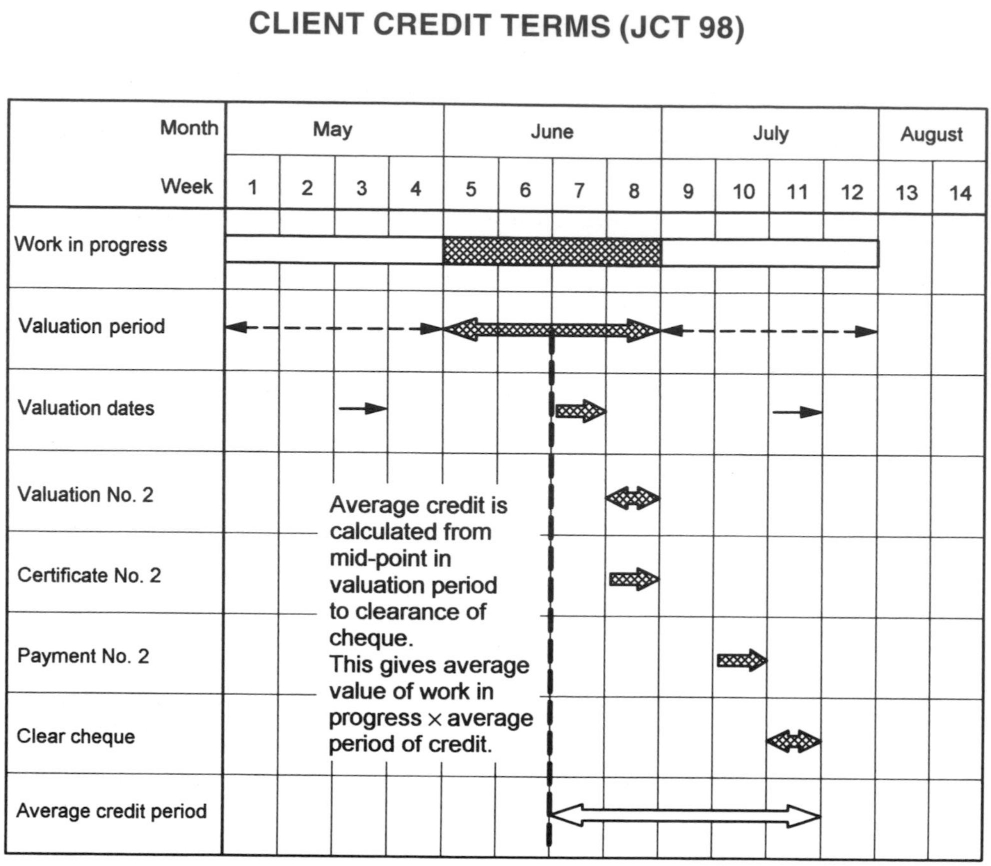

CLIENT CREDIT TERMS (JCT 98)

Month	May				June				July				August	
Week	1	2	3	4	5	6	7	8	9	10	11	12	13	14
Work in progress														
Valuation period														
Valuation dates														
Valuation No. 2														
Certificate No. 2														
Payment No. 2														
Clear cheque														
Average credit period														

Average credit is calculated from mid-point in valuation period to clearance of cheque. This gives average value of work in progress × average period of credit.

Figure 6.8

was a central theme in the Latham Interim Report 'Trust and Money'. Suppliers are commonly kept waiting for 60 or 90 days and may be forced to suspend supplies or even withdraw credit facilities from persistent offenders. Recovery action through the courts may well ensue. Some subcontractors never see their retention moneys despite years of waiting.

Typical credit terms in the industry are given in Table 6.1.

The construction industry operates a system of retentions where a percentage of the payments made for work in progress is held back by the client from the contractor (or by the contractor from subcontractors). Typically this is 3% or 5% but can be 10% or more where non-standard forms of contract are used.

The money held back is effectively capital lock-up as far as the contractor (or subcontractor) is concerned and this has a cost effect in terms of interest payments on working capital requirements. This adds to the cost of building and is counter-productive from a client value point of view.

Table 6.1 Typical credit terms.

Contract	Payment terms	Overdue payments
JCT 98 Standard Form of Contract	Monthly, 14 days from architect's certificate	Simple interest on overdue payments = Bank of England Base Rate + 5%
JCT 98 Standard Form of Contract With Contractor's Design	14 days from contractor's application	Simple interest on overdue payments = Bank of England Base Rate + 5%
JCT Major Project Form 2003	14 days after issue of payment advice	Simple interest on overdue payments = Bank of England Base Rate + 5%
ICE7 Conditions	28 days from contractor's statement	Compound interest at 2% over bank base lending rate
New Engineering Contract	3 weeks from project manager's assessment date	Compound interest at the rate stated in the contract data (minimum 2%)
JCT Standard Form of Domestic Sub-Contract	21 days after the appropriate interim certificate for the main contract is payable	Simple interest on overdue payments = Bank of England Base Rate + 5%
Civil Engineering Contractors Association Form (Blue Form)	38 days after the main contractor's application to the employer	Interest at the same rate specified in the main contract
Non-standard forms	Varies enormously but between 50 and 70 days from invoice is not unusual	Statutory rate applying at the time of court judgement
Typical plant hire contract	15 or 30 days from invoice. Invoice is usually submitted when plant is off-hired	Statutory rate applying at the time of court judgement
Typical suppliers' terms (e.g. builders merchants)	30 days from end of month of delivery of materials	Statutory rate applying at the time of court judgement

The practice of retentions has been widely criticised over the years, including by Banwell (1964) and Latham (1994), and at the time of writing is under review once again.

6.8 Forecasting the contractor's income

Figure 6.9 shows the principles of assessing the payments to be made to the contractor at the end of each payment period. The final payment is released to the contractor at the end of the defects period. The histogram, shown below the line graph, allows the monthly payment sums to be more clearly presented.

The forecast of income from the contract valuations and the release of the interim payment certificate to the contractor depend on the payment terms contained in the various forms of contract.

PRINCIPLES OF FORECASTING MONTHLY PAYMENTS DUE TO CONTRACTOR

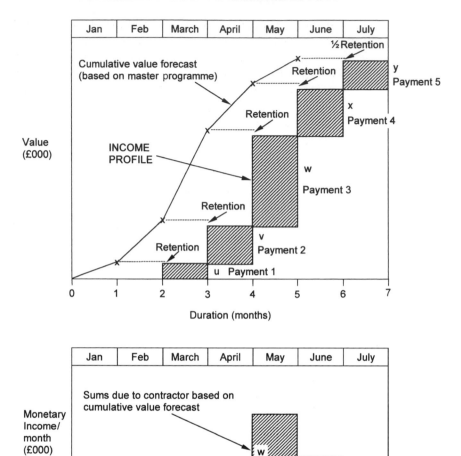

Figure 6.9

JCT 98 contract terms

Under JCT 98 conditions, the payment terms are stated as 14 days from the date of issue of the interim certificate. The overall period, however, between the date of the interim valuation and the payment of the money into the contractor's bank account, may cover some 28 days (see Figure 6.8). This time delay is made up of:

- 7 days between the valuation date and the issue of the interim certificate

- 14 days payment period
- 7 days to receive and clear the payment

The interim payment is subject to retention, as stated in the contract appendix, which may be 3% or 5% depending on the overall value of the project. The retention is reduced by one half on issue of the practical completion certificate, and the balance of the retention fund is released on issue of the certificate of making good defects at the end of the defects liability period. The defects liability period is normally 6 months for building work and 12 months where there are significant mechanical engineering installations.

WCD 98 (With Contractor's Design) contract terms

Under the WCD 98 design and build conditions, the payment terms may be based on periodic payments (normally monthly) or on agreed stage payments or milestones. The payment period from the payment request date is specified in the contract as 14 days. Similar terms to JCT 98 apply to the issues of retention and defects liability.

ICE Conditions of Contract (7th Edition)

For civil engineering contracts where the 7th edition is used, valuations are agreed between the engineer and the contractor and payment becomes due within 28 days of the issue of the certificate.

Worked example

The cumulative value forecast shown in Table 6.2 has been obtained from a master programme.

Table 6.2 Cumulative value forecast.

Month	Cumulative value forecast (£)	
1	45 000	
2	124 000	
3	198 000	Payments = monthly
4	265 000	Retention = 3%
5	320 000	
6	380 000	

FORECAST OF CONTRACTOR'S MONTHLY PAYMENTS

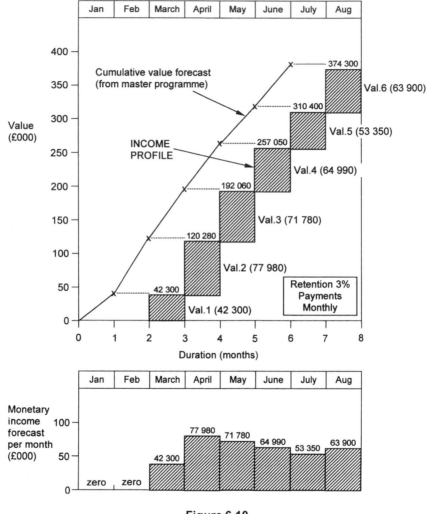

Figure 6.10

Figure 6.10 shows the cumulative value forecast presented graphically, together with the forecast cumulative and monthly income profile.

6.9 Labour, plant and preliminaries budgets

Budget forecasts may be developed for labour, plant and preliminaries expenditure based on an analysis of the contractor's net estimate, in other words excluding profit and overheads.

Moneys may be allocated to time on a bar chart and extended into a graphical display in the form of a value–time curve forecast. During the project the contractor may then monitor actual expenditure and use the analysis as part of his monthly cost control and reporting procedures.

Labour budget

Figure 6.11 indicates a bar chart display for labour expenditure. The graphical display in Figure 6.12 has been presented as a cumulative forecast of man-weeks expenditure plotted against time. This allows the actual man-weeks expended on the project to be matched with the forecast as the project proceeds.

This simple approach is highly effective as a cost control aid, but needs to be reconciled with the contract progress position in order to be an effective management tool. The reason for the variance of two man-weeks could be due to the contract being behind programme. This could have resulted from insufficient labour being available to the contractor.

Plant budget

Figure 6.13 shows a plant budget for a project in bar chart form, and Figure 6.14 the graphical display of the forecast cumulative plant expenditure.

Plant costs on site frequently exceed expenditure allowances in the tender, and site managers are notorious for keeping plant on site for too long just 'for convenience'. Whether plant is hired or owned by the contractor, strict control over budget variances needs to be exercised and plant and equipment must be off-hired as soon as it is not required.

Similar principles may be applied to the preparation of a contract preliminaries budget by analysing the preliminaries allowances in the estimate, preparing a bar chart, and allocating moneys to time.

The application of cost reporting at monthly intervals forms part of the discussion in Chapter 9.

Preliminaries budget

One of the most common areas of overspending on a project is on contract preliminaries. This may be due to management failing to allow sufficient moneys to cover this section of the tender or over-elaborate site organisation. It is also the area most subject to adjustment at the tender adjudication stage, either to make the tender more competitive or because the contractor decides to tender on a shorter programme period. The total of preliminaries typically represents between 5% and

LABOUR EXPENDITURE BUDGET

OPERATION	Labour budget man hours	Man weeks	1	2	3	4	5	6	7	8	9	10	11	12	13	14	15	16	17
Foundations	800	20	4	4	4	4	4												
Brickwork to DPC	360	9					4	5											
Ground floor Slab	200	5							5										
External brickwork	960	24							2	6	6	6	4						
Roof		S/C																	
1st. fix	240	6													2	2	2		
Plaster		S/C																	
Weekly man weeks			4	4	4	4	8	5	7	6	6	6	4	0	2	2	2	0	0
Cumulative man weeks			4	8	12	16	24	29	36	42	48	54	58	58	60	62	64	64	64

Man weeks

Figure 6.11

LABOUR BUDGET

CUMULATIVE FORECAST MAN WEEKS

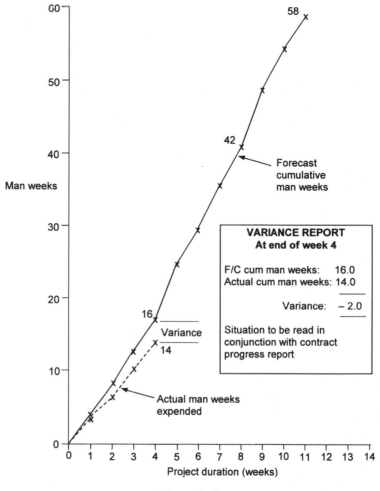

Figure 6.12

15% of the contract sum but on small contracts the percentage could be a lot higher.

Figure 6.15 indicates a tabular display for a preliminaries budget. This has been prepared in bar chart format with the preliminaries expenditure forecast presented to a time-scale. Fixed costs relating to the establishment of the site and its consequent dismantling are also indicated. Figure 6.16 shows the cumulative expenditure forecast presented as a line graph.

Actual preliminaries costs may be matched with the forecast expenditure in tabular and graphical form as the contract proceeds, and in this

PLANT EXPENDITURE BUDGET

OPERATION	Plant allocated	Budget £	1	2	3	4	5	6	7	8	9	10	11	12	13	14	15	16
Site clearance	P.Shovel 4 Lorries	2000	400	800	800													
Excavate foundations	Hyd. B/A. 2 Lorries	5000			1000	1000	1000	1000	1000									
Concrete foundations	Concrete pump	1000								1000								
Brickwork DPC	Mixer	200									100	100						
Ground floor slab	Concrete pump	1000											1000					
External walls	Scaffold mixer	4000												1000	1000	1000	1000	
1st Fix	Crane	1000																1000
Weekly expenditure forecast			400	800	1800	1000	1000	1000	1000	1000	100	100	1000	1000	1000	1000	1000	1000
Cumulative expenditure forecast			400	1200	3000	4000	5000	6000	7000	8000	8100	8200	9200	10200	11200	12200	13200	14200

Figure 6.13

PLANT BUDGET

CUMULATIVE EXPENDITURE FORECAST

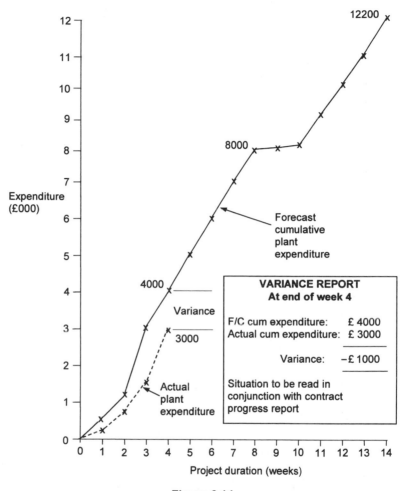

Figure 6.14

way the preliminaries variance may be monitored continuously during the project. The preliminaries budget is an ideal application for a spreadsheet.

6.10 Cash flow

The construction industry has a bad reputation for its high levels of corporate and individual insolvencies, especially among contractors. Surprisingly, many of the companies and sole traders that get into

CONTRACT – BOLTON FACTORY
CONTRACT VALUE – £600 000

PRELIMINARIES BUDGET

BUDGET FIGURES ARE NET PRELIMINARIES SUMS IN THE BILL

OPERATION	BUDGET £	1	2	3	4	5	6	7	8	9	10	11	12	13	14	15	16	17	18	19	20	
Establish site	8 000	4	4																			
Site accommodation																						
Establish	1 000		1																			
Hire	200/w			0.2	0.2	0.2	0.2	0.2	0.2	0.2	0.2	0.2	0.2	0.2	0.2	0.2	0.2	0.2	0.2	0.2		
Dismantle	1 000																				1	
Site management																						
Site manager	1 000/w	1	1	1	1	1	1	1	1	1	1	1	1	1	1	1	1	1	1	1	1	
General foreman	500/w								0.5	0.5	0.5	0.5	0.5	0.5	0.5	0.5						
Site engineer	500/w	0.5	0.5												0.5	0.5						
Temporary roads	2 000	0.5	0.5					0.5	0.5													
Hardstandings	2 000			1	1																	
Site hoarding																						
Erect/dismantle				1	1															1	1	
Weekly expenditure forecast		6	7	3.2	3.2	1.2	1.2	1.7	2.2	1.7	1.7	1.7	1.7	1.7	2.2	2.2	1.2	1.2	1.2	2.2	3	
Cumulative expenditure forecast		6	13	16.2	19.4	20.6	21.8	23.5	25.7	27.4	29.1	30.8	32.5	34.2	36.4	38.6	39.8	41	42.2	44.4	47.4	
Actual weekly cost		5	5	2	2	2	2	2	2													
Cumulative weekly cost		5	10	12	14	16	18	20	22													

Time in weeks

Total prelims £47 400

Figure 6.15

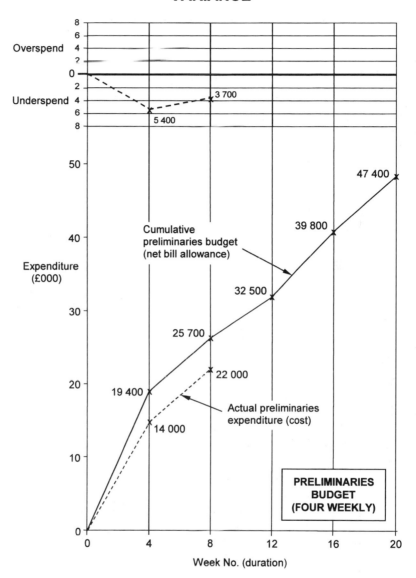

Figure 6.16

financial difficulties are generating satisfactory profits but fail because they simply run out of cash. An insolvent company is unable to pay its debts as and when they fall due and, if cash is not available at the right time, the company ceases trading.

In the case of *Gilbert Ash (Northern) Ltd* v. *Modern Engineering (Bristol) Ltd* (1973), Lord Denning, Master of the Rolls, famously said that 'there must be cash flow in the building trade – it is the very lifeblood of the enterprise'. However, cash flow is a problem not only for contractors but also for developers and construction client organisations.

Cash flow may be defined as 'the movement of money in and out of the firm'. Consequently, payments made for contract work represent negative cash flow (money out) for the client and positive cash flow (money in) for the contractor.

Client's cash flow

Developers and other clients to the industry provide the capital investment required for construction projects to go ahead. This money may either be borrowed from banks, provided by shareholders' investments or generated from profits, or a combination of all three.

The developer or client has a different view of cash flow from the contractor because their cash position is always negative until sales income or revenue from the completed building is forthcoming. An example of a client's cash flow is given in Table 6.3.

The cash flow position of two types of client is shown in Figure 6.17, which shows the cash requirements of a developer client building three houses and a client building a speculative office block project.

Contractor's cash flow

The contractor's cash flow position is somewhat different from that of the client. The contractor relies on interim or stage payments from the client to provide money in and this helps to pay for the money out payments for wages, materials, subcontractors, etc. However, because the contractor has to wait for perhaps 2 or 3 months for his money to come in, reliance also has to be placed on the credit provided by suppliers and subcontractors in order to reduce the negative cash flow effect of contract payments. An example of a contractor's cash flow is given in Table 6.4.

Table 6.3 Client's cash flow.

Money in	Money out
Housing development	Land purchase
Deposits	Interest on borrowings
Sales completions	Planning and legal fees
Rental income	Professional fees
Production revenues	Infrastructure costs (e.g. roads and sewers)
Sales of completed buildings (e.g.	Site remediation (e.g. removal of contamination)
speculative offices and factories)	Building costs (monthly or stage payments)

CLIENT CASH FLOW

Small housing development project

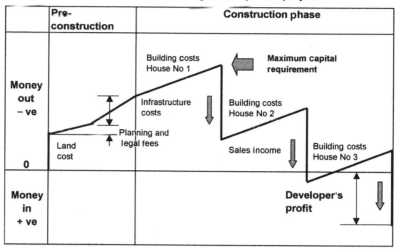

4-month office block development

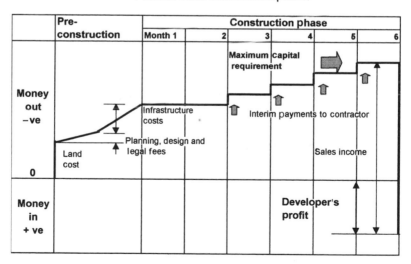

Figure 6.17

6.11 Movement of money

The pattern of money-in and money-out on a contract is illustrated as a saw tooth diagram in the top part of Figure 6.18, which shows that the contractor is in a negative cash position until month 6.

Table 6.4 Contractor's cash flow.

Money in	Money out
Monthly payments on contracts	Head office running costs
Final account payments	Staff salaries
Retentions released on practical	Company cars and expenses
completion	Payments to suppliers
Retentions released on issue of the final	Payments for plant hire
certificate	Contract payments to subcontractors
Returns on investments	Wages and labour-only payments
(e.g. land and property)	
Medium–long term bank borrowings	
Shareholders' funds invested in the business	

In the first 2 months of a typical project, the contractor will be paying out money for wages, materials, subcontractors, etc. and this represents negative cash for the contractor. Some of this money will be recovered when the client pays the first interim certificate at the end of month 2. This is unlikely to be sufficient to put the contractor in the black, however, because by this time another month's work will have been done and more costs will have been incurred.

A further negative cash flow consideration is that the client will keep some retention money back from the contractor to cover for possible defects in the work. It can therefore be seen that the contractor is in a negative cash position for a considerable period of the contract, until such time as the contract payments begin to outweigh the moneys being spent.

Figure 6.18 illustrates the contractor's cash position assuming that there is no delay in paying for the wages, materials and services required for a contract. Realistically, however, the contractor will use the credit facilities offered by his suppliers and subcontractors to offset the negative cash flow effect of having to wait for the client's payments. The payment delay situation is shown in Figure 6.19.

Harris and McCaffer (2000) suggest that a weighted average payment delay may be calculated to demonstrate the effect of using credit facilities to improve cash flow. An example of this is given in Table 6.5.

Because most construction work is carried out on credit, a delicate balance exists in the supply chain between income and expenditure. For instance, under the standard forms of contract (e.g. JCT 98 and ICE7), the contractor works for a month but the client is not required to pay for work in progress until a valuation has been carried out and a certificate issued by the architect or engineer. This means that the contractor will have to wait for a month or more before getting paid and he will have to find the money to pay wages and salaries in the meantime.

On the other hand, the contractor does not pay straightaway for sub-contract services, plant hire or materials supplied until a 'credit' period

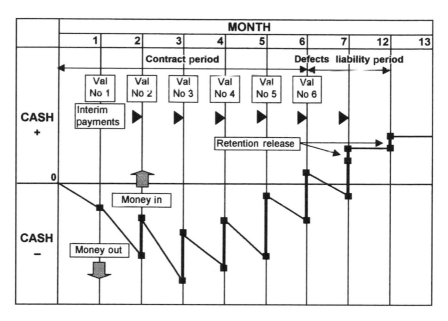

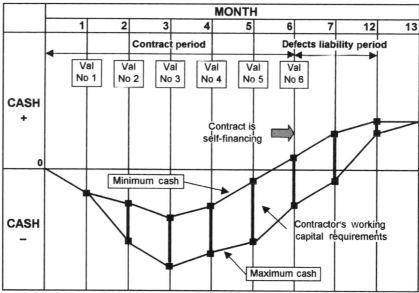

CASH FLOW PRINCIPLES
(no payment delay situation)

Figure 6.18

has expired. Subcontractors will be paid according to the terms of their contract (2–4 weeks) and suppliers perhaps 30 days from the end of the month when the materials were delivered. Plant hire companies usually invoice when the plant has been off-hired from site, and payment is

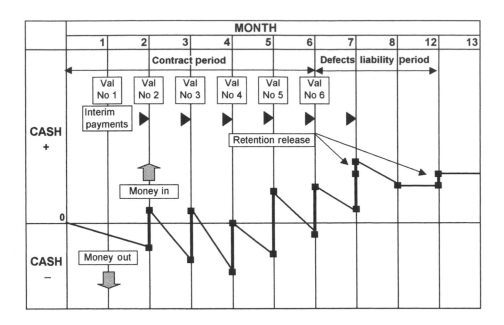

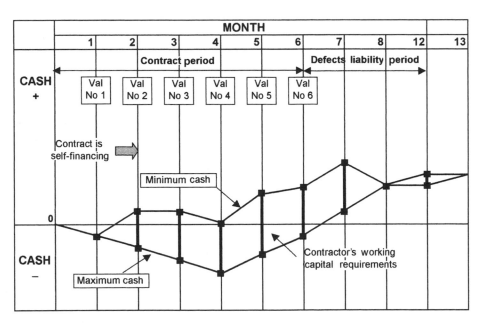

CASH FLOW PRINCIPLES
(payment delay situation)

Figure 6.19

Table 6.5 Weighted average payment delay.

	% of contract value	Payment delay or credit period (weeks)	Weighted average payment delay (weeks)
Labour	10	1	0.1
Materials	20	6	1.2
Plant	20	6	1.2
Subcontractors	50	3	1.5
Total	**100**		**4.0**

normally due a month later (see earlier in this chapter under Credit Terms in Construction).

The credit system works well unless the chain is broken and one party or the other fails to pay on time. When this happens, cash flow problems can arise for contractors or subcontractors with inadequate working capital. The longer the contractor can defer paying his creditors, the better it will be for his cash position, but care needs to be exercised to avoid withdrawal of credit facilities or possibly court action for recovery of moneys outstanding.

Creditors cannot be paid out of profits if the cash is not available and so delaying payment to creditors, while awaiting payment on contracts, has become an art form in construction. However, when small firms are squeezed between a large debtor (e.g. a main contractor) and a creditor (e.g. a supplier wanting payment), with one controlling the money and the other controlling credit facilities for materials, serious problems can occur. The Housing Grants, Construction and Regeneration Act 1996 was enacted to address these problems but in practice payment problems are still common in the industry.

6.12 Working capital

Working capital is the money a firm needs to fund its day-to-day operations in order to pay bills on time.

In order to fund the negative cash situation, the contractor needs access to working capital in the form of shareholders' funds, bank borrowings or overdraft facilities. Contractors who rely solely on overdrafts rather than long-term funding are vulnerable to insolvency because overdrafts can be withdrawn without notice.

The second diagram in Figure 6.18 illustrates the contractor's cash position and working capital requirements. The area between the zero cash line and the minimum cash line indicates the least amount of working capital needed by the contractor assuming that contract payments are received on time. If this is not the case, the maximum cash line would apply and the area between the zero cash position and maximum cash line would represent the working capital needed to fund the project.

The areas between the zero cash line and the minimum and maximum cash lines represent the contractor's minimum and maximum capital lock-up on the contract. It is this area which needs to be funded out of working capital.

Figure 6.19 shows a payment delay of 1 month (4 weeks) which means that the contractor is in the black by month 2 rather than month 6. As a rule of thumb, a 1-month payment delay improves cash flow by approximately 50% with a consequent reduction in the contractor's working capital requirements.

6.13 Improving cash flow

Despite relatively low interest rates at the time of writing, the cost of borrowing is still a matter of concern for contractors because profit margins are so low and banks lend at a premium over the base rate. Consider the situation for a contractor with an annual turnover of £20 million, annual profits of £600 000 and bank borrowings of £3 million with a lending rate of 2% over base:

Actual cost of money = base rate (say) 5% + 2% = 7%

Cost of borrowings = 7% × £3 000 000 = £210 000

Cost of borrowing expressed as a percentage of turnover = 1%

With profits at only 3% of turnover, the contractor is vulnerable should interest rates rise or profits fall. Profit in contracting is always at risk because of the uncertain nature of the work and the likelihood of mistakes in tenders and inaccurate forecasts on contracts. For these reasons, and because contractors invariably make more money by keeping other people's money for as long as possible, new ways of reducing negative cash flows are always attractive. Some common methods are listed below.

At tender stage

These methods will bring in early money but must be done before submitting the priced bills:

- Load money into undermeasured items
- Load money into early items such as excavation and substructures
- Load money into mobilisation items in the preliminaries

During the contract

These methods will reduce working capital requirements:

- Submit interim application on time
- Overmeasure the work in progress
- Overvalue materials on site
- Agree the value of variations as soon as possible
- Keep good records and submit claims early
- Deal with defective work quickly to avoid delayed payment
- Make maximum use of trade credit facilities

Post-contract

These methods will increase profit levels:

- Submit all documentation as soon as possible
- Ensure timely release of retentions by submitting health and safety file information on time
- Agree final account as soon as possible
- Collect outstanding retentions on time

6.14 Forecast value and cash flow example

Project brief

A contractor has obtained a contract for a project to be undertaken in three phases. Details of the phases with approximate durations and monetary values are shown in Table 6.6. The overall project duration is 24 months.

Project task

Produce a cumulative value forecast for the complete project and an assessment of the contractor's working capital requirements for the first 6 months of the project period. This is to be based on the following:

- The above values include a 5% contribution to profit and overheads
- 5% retention is to be applied to the payments
- Costs are to be paid at the end of the month in which they are incurred (i.e. no delay in meeting the cost situation)
- Interim payments are to be made monthly, payable 1 month after the valuation date (JCT 98 contract)

Table 6.6 Tender analysis.

Phase 1	Two storey extension	
	Value = £1 500 000	
	Duration = 10 months	
	Operations	£
	Substructure	150 000
	Superstructure	600 000
	Finishings and services	750 000
Phase 2	Refurbished canteen	
	Value = £750 000	
	Duration = 8 months	
	Operations	
	Demolitions	37 000
	Superstructure	225 000
	Finishings and services	488 000
Phase 3	Office refurbishment	
	Value = £1 500 000	
	Duration = 10 months	
	Operations	
	Demolitions	75 000
	Superstructure	450 000
	Finishings and services	975 000
Contract preliminaries		250 000
Total project value		**4 000 000**

Worked solution

The approach to the assessment of the cumulative value forecast and the working capital requirements involves the following stages:

- *Step 1*
  Assess the cumulative value forecast for the three phases of the project by allocating project values to a bar chart programme.
  Figure 6.20 shows a bar chart display for the three phases of the project with moneys allocated to project operations. The value for contract preliminaries has been allocated separately throughout the contract period with an extra sum of £10 000 for establishing the site at week 1 of the project. The cumulative value forecast is presented along the bottom of the bar chart and displayed as a value–time graph in Figure 6.21.

- *Step 2*
  Establish the cumulative cost forecast remembering that value is cost plus margin.

$$\text{Cost} = \text{value} \times \frac{100}{(100 + \text{margin})}$$

CASH FUNDING FOR THREE PHASE PROJECT

Phase I must be complete before Phase II and Phase III start

OPERATION		VALUE (£'000)	1	2	3	4	5	6	7	8	9	10	11	12	13	14	15	16	17	18	19	20	21	22	23	24
Subst.	PH.I	150	75	75																						
Superst.		600		100	100	100	100	100	100																	
Finishings		750						150	150	150	150	150														
Demolition	PH.II	37											37													
Superst.		225												45	45	45	45	45								
Finishings		488															122	122	122	122						
Demolition	PH.III	78															78									
Superst.		450																90	90	90	90	90				
Finishings		972																			162	162	162	162	162	162
Prelims	ALL	250	20	10	10	10	10	10	10	10	10	10	10	10	10	10	10	10	10	10	10	10	10	10	10	10
Monthly value			95	185	110	110	110	260	260	160	160	160	47	55	55	55	255	267	222	222	262	262	172	172	172	172
Cumulative monthly value			95	280	390	500	610	870	1130	1290	1450	1610	1657	1712	1767	1822	2077	2344	2566	2788	3050	3312	3484	3656	3828	4000

Time in months

PHASE I · PHASE II · PHASE III

Figure 6.20

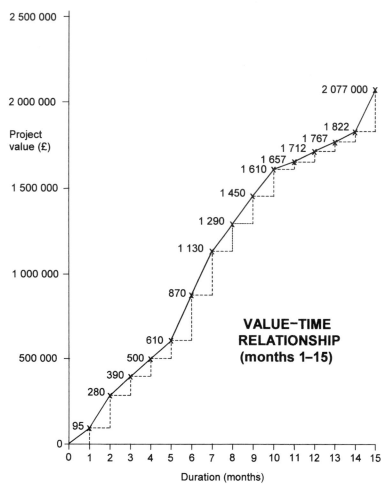

CUMULATIVE VALUE FORECAST BASED ON CONTRACT PROGRAMME

Figure 6.21

Therefore, where cumulative value is £390 000 and margin = 5%

$$\text{Cost} = £390\ 000 \times \frac{100}{(100 + 5)} = \textbf{£371 428}$$

Calculate remaining costs and plot the cumulative cost–time relationship on the graphical display. (See Figure 6.22)

- *Step 3*
  Calculate the contractor's actual income allowing for a 1-month payment delay and 5% retention.

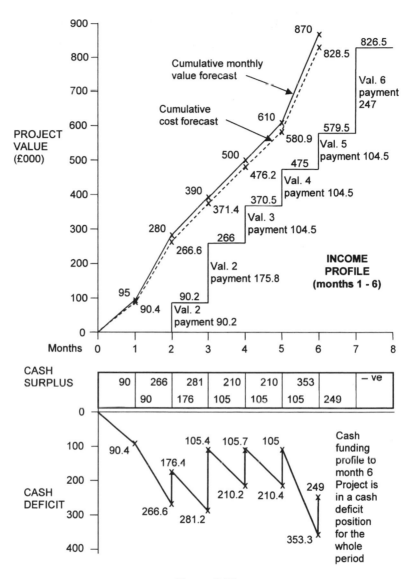

Figure 6.22

Interim payment No. 1

Forecast value	£95 000
Less 5% retention	£4 750
	£90 250

Calculate remaining payments and plot them on the display making sure to allow a 1-month delay. (See Figure 6.22)

Table 6.7 Working capital requirements.

Month	Max. cash requirement (£)
1	90 400
2	226 600
3	281 200
4	210 200
5	210 400
6	353 300

- *Step 4*
  Plot a saw tooth diagram for the first 6 months of the project using the cumulative cost and payment figures calculated above. Money out (cost) is plotted downwards at an angle and money in (payment) is plotted upwards. (See bottom of Figure 6.22)

- *Step 5*
  Display the maximum and minimum working capital requirements in the form of a table.

Figure 6.22 indicates the relationship between value–time, cost–time, and income–time for the first 6 months of the project period. The cash funding profile is shown in the form of a saw tooth diagram under the graphical display. This is based on the no delay situation in meeting the cost as stipulated in the project brief.

The cash funding profile indicates that during the first 6 months of the project period, cash requirements peak at the end of months 3 and 6 respectively. The maximum working capital requirements at the end of months 1–6 are shown in Table 6.7.

References

Banwell, Sir H. (1964) Report of the Committee (Chairman Sir Harold Banwell). *The Placing and Management of Contracts for Building and Civil Engineering Work.* HMSO.

Hall, L. (1974) *Business Administration.* M and E Handbooks, McDonald and Evans.

Harper, W. M. (1976) *Management Accounting.* M and E Handbooks, McDonald and Evans.

Harris, F. & McCaffer, R. (2000) *Modern Construction Management*, 5th edn. Blackwell Publishing.

Latham, Sir M. (1994) *Constructing the Team.* HMSO.

Pilcher, R. (1975) *Principles of Construction Management*, 2nd edn. McGraw Hill.

7 Managing the supply chain

7.1 Supply chain integration

In the main, the construction industry is concerned with one-off projects, geographically dispersed and carried out by short-term teams of designers, contractors, subcontractors and suppliers who have been assembled specifically for the project, only to be disbanded once the contract is complete.

In order to overcome this problem, some major clients have assembled their own standing lists of preferred suppliers so that their projects can be managed on a more long-term and integrated basis while retaining an appropriate degree of competition and control.

These suppliers, or supply chain partners, include architects, engineers, main contractors, specialist contractors and suppliers of a variety of goods and services appropriate to the client's projects. In the wider context of the industry at large, however, this principle tends not to penetrate the first tier of the supply chain so as to facilitate long-term relationships between contractors and *their* subcontractors and suppliers.

Consequently, one-off clients to the industry are unable to reap the benefits of supply chain integration to the same extent as repeat customers do and this leads to poor performance and client dissatisfaction.

Supply chain integration does not, however, rely entirely on successive projects for the same client to ensure collaboration and the benefits of long-term relationships and understanding.

Under the traditional system of contracting, architects and other designers rely heavily on many specialists and suppliers for design information and advice. However, unless these firms are nominated under the contract, the main contractor is likely to choose different companies to do the work once the contract has been awarded. This is not conducive to efficient working.

Informed thinking suggest that smaller and more integrated supply chains are the answer to demands from clients for increased productivity from the industry and greater regard to completion on time, budget certainty and higher standards of quality and health and safety management. To this end, many of the top firms in the industry have changed their culture entirely by having:

- Fewer and better clients
- Fewer and better suppliers and subcontractors

Large contractors may have between 5000 and 15 000 suppliers of various sorts:

- Specialist subcontractors
- Material suppliers
- Builders merchants
- Plant hire firms
- Hire firms for small tools
- Office equipment suppliers
- Stationery suppliers
- Computer equipment and software suppliers
- Domestic firms of cleaners
- Suppliers of cleaning equipment
- Suppliers of tea, coffee and milk

The reasons for such large networks of suppliers lie in the outmoded belief that this is the way to encourage competition and thereby lower prices. The effect is quite the contrary and a great deal of time and effort is needed to obtain competitive quotations from two, three or six suppliers every time supplies or services are required.

By reducing the number of supply chain members and by entering into long-term relationships with them, many time-consuming and costly formalities can be avoided, thus adding value to the supply chain.

Therefore, instead of following the procurement procedures outlined in this chapter for, say, 6 subcontractors for each of 30 or more trades in a typical contract, the contractor can follow the same procedures with fewer subcontractors. The time saved can then be spent discussing long-term price frameworks, delivery schedules, construction programmes and health and safety and quality issues.

One of the touchy aspects of supply integration is the need to assure profits. These clearly need to be ring-fenced but many suppliers and subcontractors might be wary of revealing too much information to the contractors with whom they do business. However, in order to follow the Toyota model referred to in the Egan Report, the process of team working and integration must penetrate the entire supply chain in order to reap the available benefits.

Many examples of supply chain integration may be taken from practice, such as the major client who is constructing 20 motel complexes each year at an annual cost of £30 million. The client is able to work out a deal with a preferred contractor and both the client and contractor can integrate major suppliers and specialists into the project programme at an early stage.

Consequently, construction lead times are vastly reduced and just-in-time delivery can be assured for items such as the timber frame package, plumbing and electrical fit-out packages, the supply of fitted out bathroom and bedroom pods, etc. The whole supply chain is assured of continuity of work and the contractor assured of his ring-fenced profit margin.

7.2 Lean construction

One of the great benefits of an integrated supply chain is the opportunity to apply lean thinking to the construction process. The Egan Report 'Rethinking Construction' (Construction Task Force 1998) advocated this principle which Cartlidge (2002) defines as 'the elimination of waste from the production cycle'. Cartlidge explains that *every time waste is removed from the supply chain, value is added to the process, leading to lower costs, shorter construction periods and greater profits.*

Part of the lean thinking approach is the use of just-in-time production where materials and components are manufactured, transported and delivered to site as and when required without the need for long lead-times and stockpiling on site.

'Rethinking Construction' describes lean production as a generic version of the Toyota Production System, which is recognised as the most efficient in the world. The system is based on lean thinking principles including:

- Elimination of non-value activities which can represent up to 95% of time and effort
- Removal of waste from all activities involved in delivering the end product
- Establishment of relationships with all members of the supply chain
- Removal of delays in the design and production process using just-in-time management

Lean thinking can be successfully applied to the construction process through innovative design and assembly, including the use of off-site manufacture, pre-fabrication, pre-assembly and supply chain integration.

7.3 Fast track construction

Many clients to the construction industry now demand early delivery of their project requirements. This is especially the case with commercial clients who are looking for shorter and shorter construction periods so as to ensure that the completed facility is on line and earning revenue at the earliest possible moment. As a consequence of this demand, procurement methods which overlap design, tendering and construction have emerged, but alongside these developments other approaches to the management of construction projects have developed including fast track construction.

Kwakye (1997) describes the fast-track system as a management approach aimed at the early completion of the construction phase using a combination of innovative procurement methods, industrialisation of the construction process and the use of work package contractors in order to benefit from their expertise, especially as regards their design input. The benefits of the system include:

- Overlapping of work packages both during design and construction
- Less duplication of effort and waste
- Less uncertainty and inefficiency at work package interfaces
- The use of innovative construction methods
- Incorporation of cutting edge technologies
- More emphasis on the standardisation, pre-assembly and modularisation of the construction process

As a consequence, the fast track system has the propensity to deliver completed projects in remarkable time-scales. The result is much faster on-site construction periods such as a new superstore in 11 weeks and a new fast food outlet in 2 weeks.

The intensive construction programme which results from this approach requires high standards of planning, organisation and control. It is essential to ensure that everything is right first time and that there are zero accidents so as to prevent unnecessary delays and disruption to the programme.

Fast tracking of projects does not happen by accident and a great deal of planning and preparation is necessary to create the right client–supplier relationships. These are characterised by open and honest dealing, encouragement of innovation and the use of performance specifications in order to:

- Improve buildability by involving contractors in the design phase
- Involve first tier suppliers and works package contractors at an earlier stage
- Reduce construction lead times through pre-ordering of key materials and components (e.g. structural steel, timber frame, concrete and cladding, partitions, ductwork, air handling units, modular components, etc.
- Arrange the direct supply of materials and components where

repeat clients can benefit from volume discounts and higher quality by dealing direct with key suppliers

7.4 Subcontractors

There are very few general contractors around in the construction business these days as most projects rely on a managing contractor to organise and supervise the work and a series of specialist subcontractors who carry out the site operations.

Subcontracting is the vicarious performance of a contractual obligation where:

- The main contractor normally enters into a direct contract with each subcontractor
- Subcontractors may sublet their work to sub-subcontractors
- The main contractor retains contractual responsibility for the subcontractor's work
- Subcontractors are commonly referred to as:
  o Work package contractor
  o Trade contractor
  o Specialist contractor
- Attendances are normally provided by the main contractor, which may include the provision of water and power, use of standing scaffolding, use of tower crane and removal of rubbish
- Special attendances may be supplied, such as offloading of materials, provision of hardstandings, task lighting, etc.

7.5 Types of subcontractor – traditional procurement

Domestic subcontractor

- A subcontractor chosen and engaged directly by the main contractor to carry out a particular trade
- May be labour-only or supply and fix
- Examples – groundwork, brickwork, roof tiling, plastering. Usually appointed under a standard form of contract or using the main contractor's bespoke conditions. Under JCT 98, the contractor requires the architect's approval to sublet any part of the works but approval of particular subcontractors is not required
- The ICE Conditions require the contractor to notify the engineer of the extent of work to be subcontracted and the names and addresses of each subcontractor. The engineer may object to the employment of any subcontractor

Nominated subcontractor

- A subcontractor chosen by the architect (JCT 98) or engineer (ICE7) to carry out specialist work
- Appointed on the expenditure of a provisional sum or prime cost sum included in the contract bills
- The nominated subcontractor enters into a contract with the main contractor but there may be a collateral warranty between the nominated subcontractor and the client. Nomination is not possible under the JCT Intermediate Form or the Engineering and Construction Contract

Named or listed subcontractor

- Under the JCT 98 Standard or Intermediate Forms, the architect may provide the contractor with a list of at least three subcontractors to choose from for carrying out particular parts of the project
- The contractor can add to the list but, in any event, the subcontractor is a domestic subcontractor under the contract.

Work package contractor

- Often used as a synonym for domestic subcontractor but might also imply responsibility for a specific section or package of the works, possibly including design work
- May provide a complete service with little or no reliance on main contractor attendances

Labour-only subcontractor

- Self-employed individuals, partnerships or small firms
- Higher earnings causing disparities with other workers
- Lower tax and national insurance contributions
- Lower prices but higher overheads for the main contractor
- Lower employment costs for the main contractor (e.g. no holidays with pay, redundancy and sick pay)
- Higher output and faster work leading to quality problems
- Higher materials wastage
- Lack of training and competence testing
- Supplement skills shortages in the industry
- Lower standards of safety

7.6 Types of subcontractor – management procurement

Trade contractors

- The contractors who carry out the work under the JCT Construction Management Trade Contract
- Engaged and paid directly by the client

Works contractors

- The subcontractors responsible for carrying out the works under the JCT Management Contract
- Engaged and paid directly by the management contractor

7.7 The control and coordination of subcontractors

The planning stages involved in the control and coordination of sub-contractors follow similar stages to those that the main contractor goes through, i.e. pre-tender, pre-contract and contract planning . The decision-making process in respect to subcontractors and their involvement in the project will therefore be dealt with under these headings. Subcontractor input into the various planning stages is illustrated in Figures 7.1 and 7.2.

The pre-tender planning stage

Assuming a traditional JCT 98 main contract, the subcontractor selection process follows the following stages:

- Stage 1 – Qualification
- Stage 2 – Compilation of tender list
- Stage 3 – Tender information and submission
- Stage 4 – Tender assessment
- Stage 5 – Tender acceptance

Each of the above stages will be dealt with in the form of a check-list as outlined in the CIB Code of Practice for the Selection of Subcontractors, and recommendations in the CIOB Code of Estimating Practice (1997).

Stage 1 – Qualification

Check-list for subcontractor selection:

SUBCONTRACTOR INPUT INTO THE PLANNING PROCESS

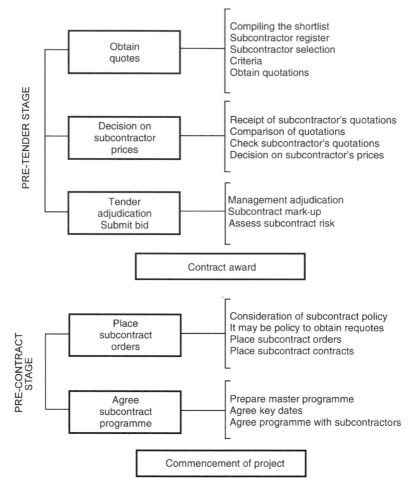

Figure 7.1

- Previous experience with the subcontractor
- The subcontractor's ability to manage his resources and liaise with the main contractor's staff. Good relationships between parties are an essential requirement to developing a team approach to a successful project
- Financial standing of the subcontractor. His ability to wait to be paid
- The expertise that the subcontractor can bring to the project
- The subcontractor's reputation and his standing with the client
- The current commitment of the subcontract organisation. Their current workload with other contractors should be determined and serious consideration given to their ability to cope with the increased

SUBCONTRACTOR INPUT INTO THE PLANNING PROCESS

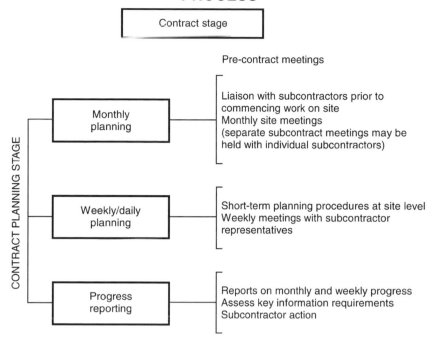

Figure 7.2

work. A large number of subcontractors just cannot say no when it comes to taking on more work. They often pull and push their limited labour force between sites hoping the main contractor will not notice that they are stretched to the limit

- The acceptability of the subcontractor to the client. On many contracts the contractor is required to name his subcontractors at the tender stage
- The competitiveness of the subcontractor's price. The price must be right otherwise the subcontractor will never win any work. Price discounts which may be applicable and the subcontractor's response to negotiation may be an important factor
- The contractual risk which the main contractor takes on the subcontract item. Low-risk subcontract operations may be let to the more risky subcontractor and hence the main contractor may include a lower subcontract price in his estimate. Subcontract operations which are critical to the success of the contract require careful consideration. For example, a high price-risk subcontract operation could be external brickwork, while site demolition work could be low risk
- The ability of the subcontract organisation to meet quality assurance criteria as laid down by the main contractor or as specified by the client

Table 7.1 Frequency of criteria use for selecting subcontractors.

Criteria	Never	Rarely	Sometimes	Always
Financial strength	6	8	12	7
Previous experience	0	0	8	25
Ability to submit a bona fide bid	1	4	16	12
Labour resources	0	2	14	17
Management capability	0	30	3	0
Current and anticipated workload	1	1	23	8
Quality of workmanship	0	2	12	19
Transportation/project location	27	4	0	2
Safety records/working practices	0	11	7	15
Reliability and trustworthiness	0	0	8	25

- References available from the subcontractor, including trade and bank references. The willingness of the subcontractor to allow previous contract work to be inspected

It is important that good relationships are established between the main contractor and subcontractor as early as possible in the planning process. This is especially important where the main contractor intends to sublet all the work on a particular project.

A survey of 33 refurbishment contractors undertaken by Okoroh and Torrence (1992) indicated that most contractors did use some form of criteria when selecting subcontractors. Table 7.1 indicates the responses to a questionnaire on the frequency with which certain selection and appointment criteria were applied. Previous experience and reliability ranked first with regard to subcontractor selection, closely followed by quality of workmanship and labour resource availability.

In the process of choosing a subcontractor, it was found to be very important to select a firm that could liaise amicably with both the contractor's head office and site staff and satisfy different interests. A further significant consideration was the acceptability of subcontractors to clients and their consultants.

Stage 2 – Compilation of tender list

As soon as it has been decided to select a subcontractor by competitive tender, a shortlist of those considered suitable to be invited to tender should be compiled. It is often the case that the main contractor has a list of subcontractors of established skill, integrity, responsibility and proven competence. This should be considered as a matter of policy within the contractor's organisation.

Main contractors should review their lists periodically so as to exclude firms whose performance has been unsatisfactory and to allow the introduction of new subcontract firms.

In some cases, the main contractor may find difficulty in compiling a short list due to the shabby way he has treated his subcontractors in the past. In this case he may have to revert to the pin-sticking approach using the building section in Yellow Pages in order to obtain quotations. There are dangers in this practice and it is much more preferable to cultivate good subcontractor relationships as recommended in the Latham Report. Many contractors are now benefiting from the wisdom of this approach.

The cost of preparing tenders is a significant element of the overheads both of the main contractor and his subcontractors. Tender lists therefore should be kept as short as practicable. Enquiries should be kept to between three and six depending on the type of subcontract work and size of project.

Perhaps the main contractor should consider adopting a policy of creating good relationships with a small group of reliable subcontractors whose businesses can expand as the main contractor becomes more established. Often the main contractor's reputation may rely solely on the excellence of his subcontractors' performance.

The use of a register of subcontractors at the selection stage is recommended by Canter (1993). This may lead to establishing subcontractor record cards relating to subcontractor performance criteria on previous contracts. It is important that such records are regularly updated and consideration given to adding new subcontractors to the register. It may be advisable at this stage to consider using a computer database, such as Microsoft Cardfile, for compiling the subcontract register. This may prove advantageous to a medium-sized contracting organisation in order to speedily access subcontract information.

It is important that close relationships are encouraged between the main contractor and his subcontractors for the survival of both parties in the long term.

Stage 3 – Tender information and submission

Both the CIB Code of Practice (CIB 1997) and CIOB Code of Estimating Practice (CIOB 1997) identify what information should be contained in the contractor's enquiry to a subcontractor. The main recommendations are summarised here as a check-list.

Details of main contract works:

- Job title and location of site
- Name of employer

- Names of architect, supervising officer, quantity surveyor and other consultants including the planning supervisor
- General description of the works

Subcontract works:

- Relevant extracts from bills of quantities and specification
- Extracts from the contract preliminaries section
- Copies of relevant drawings
- Details of where original documents may be inspected
- Time period for completion of subcontract work (if known)
- Approximate dates when subcontract work will be undertaken
- Names of adjudicator (in case of dispute)

Subcontractor's responsibility for site arrangements and facilities:

- Watching and lighting
- Storage facilities
- Unloading, hoisting and getting in materials
- Scaffolding
- Water and temporary electrical supplies
- Safety, health and welfare provisions
- Licences and permits
- Any additional facilities

Conditions of subcontract:

- Form of subcontract agreement
- Period of interim payments and payment terms including whether 'pay when paid' will apply
- Discount applicable to the payments
- Fluctuations or fixed price tender
- Other special conditions

Particulars of the main contract conditions:

- An extract from the appendix to the form of contract will assist in providing a summary of the contract particulars. This should contain the following information:
  o Form of contract
  o Fluctuations provisions
  o Method of measurement
  o Main contract period and completion date
  o Defects liability period
  o Liquidated and ascertained damages

o   Period of interim certificates
o   Basis of dayworks
o   Insurance provisions
o   Deletions or amendments to standard contract clauses

Type of quotation required from the subcontractor:

- Lump sum quotation
- Schedule of rates

Other information:

- Date for the return of the tender
- Person in the contractor's organisation to contact
- Period for which the tender is to remain open for acceptance
- Extent of the phasing of the works and number of anticipated visits to undertake the works
- Reference to any relevant attendances likely to affect the subcontractor

Similar check-list points can also be found in Canter (1993) and Brook (1993).

Stage 4 – Tender assessment

The CIOB Code of Estimating Practice (CIOB 1997) makes recommendations for analysing subcontractor quotations in the form of a domestic subcontractor register. Subcontractors' quotations are not straightforward to compare. Some do not price all the items in the enquiry, often there are mistakes and some subcontractors price net while others offer discounts, typically $2^1/2\%$. Therefore a register or spreadsheet is a useful device to enable quotations to be matched, discrepancies identified and discounts adjusted.

A typical checking procedure when comparing subcontractors' quotations should include consideration of the following:

- Does the work described in the quotation comply with the specification?
- Have all items been priced and if not are they included in other rates?
- Are unit rates consistent throughout the quotation?
- Check that the quotation does not form a counter-offer and that the subcontractor has accepted the terms and conditions of the enquiry

Stage 5 – Tender acceptance

Adjudication of the main contractor's estimate and its conversion into a tender requires management decision-making whatever the size of firm. Where the tender includes a large proportion of subcontract work, it is critical to decide carefully on the mark-up on the subcontractors' quotations selected for inclusion. It will also be necessary to carefully scrutinise the estimator's allowances for attendance on the subcontract works at the adjudication stage.

Consideration needs to be given to any late quotations received which will directly affect the competitiveness of the overall bid and this is where the subcontract comparison sheet or register facilitates last minute lump sum adjustments to be made to the tender.

The percentage mark-up on subcontractors' quotations often varies widely according to the contractor's desire to win the contract. Percentages will probably be in the range $2^1/_2$–15% depending on the contractor's view of the risk attached to the subcontract element of the contract.

Despite all the foregoing considerations, there often appears to be no rhyme or reason as to how contractors arrive at their tender adjustments.

7.8 Work package procurement

The procurement of subcontract work packages is somewhat different under management contracting and construction management contracts compared with the arrangements and negotiations on a traditional JCT 98 contract. For instance, the management contractor or construction manager has to work in conjunction with the design team's quantity surveyor in order to produce a workable budget. This may take the form of an overall budget developed from an assessment of the budget figures for individual work packages.

Deciding how to package the project is another principal difference compared to conventional contracts. Sidwell (1983) suggests a flow chart approach to the process of establishing work packages and this concept is illustrated in Figure 7.3. However, the packages cannot be thought of as subcontract enquiries in the normal sense and careful thought is needed so as to avoid too many interface problems on site. These interfaces could lead to the situation where none of the package contractors has included for a particular item of work, or duplication may occur where several firms are responsible for the same item. Either way, extra cost or disputes could result. Also, procurement of the work packages requires special negotiating skills and package contracts have to be carefully set up so as to reduce the risk of disputes.

The process involves close liaison between the design team and the appointed management contractor or construction manager in order to establish procedures for securing the best buy.

Figure 7.4 illustrates a typical work package budget for a science park project involving the construction of an office block of some 3000 m^2 in floor area. The work packages are established at the design stage and

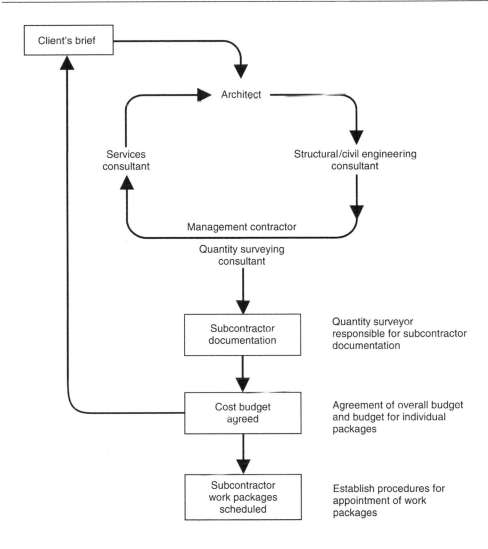

DETERMINATION OF WORK PACKAGES

Figure 7.3

budget figures have been produced from the scheme drawings based on approximate quantities and current rates.

Provision has been made on the work package budget form for entering actual work package tender prices in order to identify variances in the budget, and to update the budget as the project develops. The individual budgets may not always be achieved in practice, of course, and a swings and roundabouts approach may have to be taken while at the same time making sure the overall budget is maintained.

WORK PACKAGE BUDGET

PROJECT — UNIT ONE — WARRINGTON SCIENCE PARK				
PACKAGE REF.	WORK PACKAGE	BUDGET	ACTUAL	SAVING/ INCREASE
SUBSTRUCTURE				
100	FOUNDATIONS	110 000	100 000	–10 000
101	DRAINAGE	25 000	18 000	–7 000
102	CAR PARKS	90 000	95 000	+5 000
103	LANDSCAPING	30 000		
STRUCTURE				
200	STEEL FRAME	250 000	245 000	–5 000
201	FLOOR/ROOF DECKS	40 000		
202	ROOF FINISHES	20 000		
203	EXTERNAL ENVELOPE	220 000		
204	EXT. WINDOWS/DOORS	60 000		
FINISHES				
300	INTERNAL PARTITIONS	40 000		
301	FLOOR FINISHES	35 000		
302	CEILING FINISHES	25 000		
SERVICES				
400	HEATING	150 000		
401	ELECTRICAL	100 000		
402	AIR CONDITIONING	70 000		
403	LIFTS	80 000		
404	FIRE AND ALARM	20 000		
CONTINGENCY		50 000		
TOTALS		1 415 000		
INITIAL BUDGET ASSESSMENT		1 500 000		
FINAL BUDGET ASSESSMENT				

Figure 7.4

Many of the industry's experienced management contractors and construction managers adopt standard procedures for the selection and appointment of work package contractors and examples of these are outlined below.

Establishing work package information requirements

When the scope of the various packages has been established, it is necessary to assess lead-in times for each of them prior to commencing the procurement process. In addition, information requirements relative to each of the work packages have to be assessed and this will require the design team to work closely with each other and the management contractor/construction manager as they will have to produce the information by agreed dates prior to sending out work package enquiries.

A typical set of key dates is given below, which could be used for each of the work packages in a procurement schedule:

- Agree work package list
- Design information complete – architect/engineer/services
- Design packages to the quantity surveyor
- Schedules and quantities complete
- Tender package issued
- Enquiries sent to work package subcontractors
- Tenders returned
- Technical checks complete
- Work package negotiations complete
- Appoint work package subcontractor – letter of intent
- Arrange work package contract to be signed
- Working drawings issued
- Lead time required
- Commencement date on site planned

The establishment of key dates for each of the above items places extensive responsibility on the management contractor/construction manager. First, he has to motivate the design team to produce the information on time, clearly pointing out the consequences of failing to meet the programmed dates. Just one uncooperative member of the design team can cause the best planned project to fall apart.

The key date requirements relative to each of the work packages may be linked to the project planning software being used for the project. Many of these packages produce key date schedules relative to pre-contract activities developed from the project network diagram.

Figure 7.5 shows a work package procurement schedule for a major project. This enables planned, current and actual dates for each work package to be monitored as the packages are awarded.

PACKAGE		Issue tender documents	Collate tender documents	Out to tender	Return tender	Tender appraisal	Client review	Place order	Lead-in time (weeks)	Start date on site
100/1 Brickwork and blockwork	Planned	29/03/96	01/04/96	09/04/96	03/05/96	07/05/96	28/05/96	31/05/96	13	27/08/96
	Current	29/03/96	01/04/96	09/04/96	03/05/96	07/05/96	28/05/96	31/05/96	13	27/08/96
	Actual									
200/1 Wall cladding Excl glazing	Planned	12/04/96	15/04/96	19/04/96	17/05/96	20/05/96	03/06/96	07/06/96	17	30/09/96
	Current	12/04/96	15/04/96	19/04/96	17/05/96	20/05/96	03/06/96	07/06/96	17	30/09/96
	Actual									
200/2 Rooflights	Planned	16/02/96	19/02/96	01/03/96	29/03/96	01/04/96	15/04/96	19/04/96	16	05/08/96
	Current									
	Actual									
200/3 External glazing panels	Planned	26/01/96	29/01/96	09/02/96	22/03/96	25/03/96	15/04/96	19/04/96	24	30/09/96
	Current	26/01/96	29/01/96	09/02/96	22/03/96	25/03/96	15/04/96	19/04/96	24	30/09/96
	Actual									
300/1 Roofing and northlights	Planned	04/04/96	09/04/96	12/04/96	10/05/96	13/05/96	24/05/96	31/05/96	15	09/09/96
	Current	04/04/96	09/04/96	12/04/96	10/05/96	13/05/96	24/05/96	31/05/96	15	09/09/96
	Actual									
400/1 Plaster/ screeds	Planned	16/08/96	19/08/96	30/08/96	27/09/96	30/09/96	14/10/96	18/10/96	5	18/11/96
	Current	16/08/96	19/08/96	30/08/96	27/09/96	30/09/96	14/10/96	18/10/96	5	18/11/96
	Actual									

WORK PACKAGE SCHEDULE

Figure 7.5

Arrangements for procuring the work packages

Selection of the individual work packages often follows a standard procedure which may also be part of a quality system. It should be noted that the extent to which selection procedures are strictly adhered to may depend on the degree of risk allocated to the work package contractor.

For each of the work packages the following selection routine may be followed:

- Select eight potential work package contractors
- Shortlist six work package contractors
- Send questionnaire to each
- Financial checks
- Visit current contracts
- Reduce shortlist and interview four contractors
- Tender documents to four contractors
- Tenders received and checked
- Interview lowest two tenderers
- Appoint work package contractor in agreement with the design team

This procedure is illustrated in Figure 7.6.

7.9 Pre-contract liaison

Following the selection process, prospective subcontractors or works package contractors should be informed that their tenders have been accepted and that a contract will follow. However, before contracts are signed several matters need to be clarified so that disputes do not arise during the works. At this stage, it is useful to follow a check-list approach to ensure that all the pre-construction issues are covered. These will include:

- Confirm order or issue letter of intent
- Prepare contracts for signing
- Agree key dates, sequence of works and programme including integration with the project programme
- Agree timing for issue of drawings or approval of design information
- Confirm requirements for insurances
- Agree or obtain client/client representative approval of subcontracts or packages where there is a contractual requirement to do so
- Agree the provision of samples, sample panels and mock-ups to be submitted for client approval

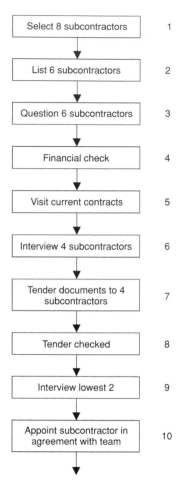

SUBCONTRACTOR/WORK PACKAGE SELECTION

Figure 7.6

- Agree the facilities to be provided by the subcontractor/package contractor, for example, cabins, stores, offloading, scaffolding, removing rubbish, etc.
- Agree dates for interim certificates and procedure for requesting payment
- Agree notification requirements for instructions, variations, dayworks, etc.
- Sign contract

Before any work commences on site, it is vital that the main contractor or the management contractor/construction manager maintains contact with the works contractor in order to keep him fully informed of the contract progress position. For instance, where there is a likelihood of the commencement of the work being delayed as much notice as possible should be given. It is essential to maintain this contact and to ensure that notification procedures agreed at the pre-contract stage are observed by all parties.

7.10 Forms of subcontract

Table 7.2 lists the main standard forms of subcontract available but there are, of course, many variations on a theme and subcontractors are frequently faced with contract conditions which they have never seen before.

Table 7.2 Standard forms of subcontract.

Name of subcontract	Main provisions
JCT Standard Form of Domestic Subcontract 2002	Used in conjunction with JCT Standard Form Provision for including procurement periods for materials and fabrication and for notice period for commencement Includes provisions for sectional completion Gives earliest and latest starting dates for the subcontract works Contractual obligation to provide information for the health and safety plan and file
JCT Trade Contract (TC/C)	Used in conjunction with JCT Construction Management Contract Commencement date and completion period stated in appendix Interim payments apply Stage payments optional
JCT Works Contract	Used in conjunction with JCT Management Contract
Civil Engineering Contractors Association Form	Used in conjunction with ICE Conditions Period for completion stated in third schedule to contract Subcontract payment conditional on main contract payment (pay when paid)
The Engineering and Construction Contract Subcontract	Used in conjunction with ECC The Accepted Programme is prepared by the subcontractor and submitted to the contractor for acceptance A method statement with resources to be provided for each operation Accepted programme to be updated by subcontractor Early warning provisions apply
ACA Standard Form of Specialist Contract for Project Partnering (SPC2000)	Used in conjunction with PPC2000 Early Warning System and Problem Solving Hierarchy to operate between constructor and specialist Specialist works to be carried out in accordance with the specialist timetable Specialist to update timetable at specified periods Acceleration provisions included in contract

7.11 Liaison during the contract period

The coordination and control of subcontractors and package contractors is crucial to the success of the project and lends itself to a standardised approach. Many of the large contracting organisations and construction managers have developed their own procedures for this. What is the secret to success? Well, by 'meetings – bloody meetings' as the John Cleese management training video recommended!

Many major contractors adopt a policy of ruling works contractors with a rod of iron, but this is not the way to ensure cooperation and team working on site. The success of a project depends on the performance and quality of those who carry out the work on site and such a short-sighted policy will ultimately fail.

Good liaison and mutual respect must be established as early in the contract period as possible and maintaining contact with subcontractors or works package contractors during the pre-contract period helps to build up an early working relationship. During the project, this is developed by regular contact at weekly and monthly progress and coordination meetings.

Figure 7.7 illustrates the meetings likely to be held during the contract stage where the works contractors will discuss problems and information requirements and iron out any difficulties which may be affecting the progress of operations on site. A check-list of points to be considered at such meetings will include:

- Review progress and quality of work
- Review the programme including relationships between the main programme operations and those of other works activities
- Action to maintain progress
- Investigate site problems and hold-ups
- Review labour situation
- Review the plant and material supply situation
- Overview of site organisation and supervision requirements
- Consider health, safety and welfare situation
- Confirm the situation regarding the issue of site instructions and variations to contract
- Review the valuation and payment situation to date

A system of short-term planning may be implemented in order to keep works contract progress under constant review. This will involve preparing weekly or 2-weekly programmes which will be discussed at the progress meetings, and preparing work plans for the next short-term period. Figure 7.8 illustrates the principles of a 2-weekly short-term planning system.

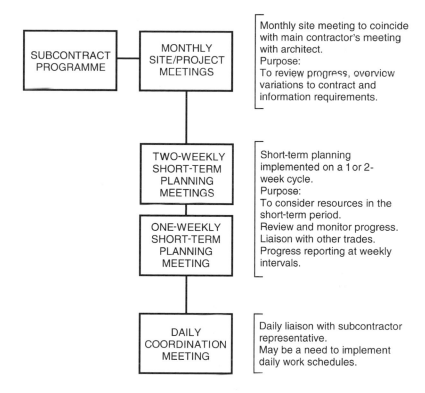

SUBCONTRACTOR LIAISON MEETINGS

Figure 7.7

It is important that the works contractors' site representatives participate in the short-term planning procedures and a good working relationship is established at site level.

With future projects in mind, it is a good idea to record the performance of the works contractors on site. This will be useful feedback for reviewing tender lists so as to ensure that works contractors are competent and adequately resourced and that there is an up-to-date database for new projects. One possible approach to this is to list the key performance criteria and to give each one a score or rating on a scale of 1 to 5. The performance criteria might be:

- Price
- Quality and workmanship
- Health and safety
- Standard of cooperation
- Time–programme performance

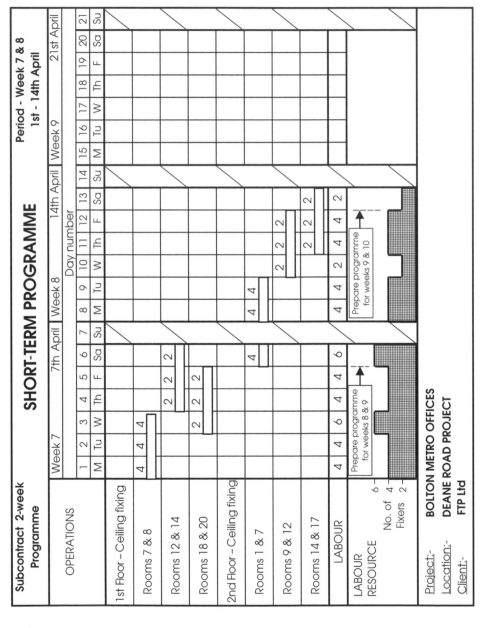

Figure 7.8

The maximum score would be 25 but a rating of, say, less than 10 might lead to exclusion from the tender list. Certain criteria, such as health and safety for instance, could be weighted for importance where appropriate.

7.12 Construction Industry Scheme

In order to remove opportunities for the avoidance of income tax and national insurance contributions, payments made by contractors to subcontractors working in construction are currently subject to Inland Revenue rules commonly known as IR35. These rules were first announced in news release No. IR35 following the 1999 Budget and are backed by statute under the Finance Act 2000 (section 60 and schedule 12).

Under the Construction Industry Scheme (CIS), subcontractors must hold either a registration card (there are five types) or a subcontractors tax certificate before they can be paid by the contractor. Subcontractors who meet certain qualifying conditions are issued with a subcontractors tax certificate by the Inland Revenue and those who do not are issued with registration cards.

In cases where the subcontractor holds a subcontractors tax certificate, the contractor will pay gross (i.e. with no deductions). However, where a subcontractor only holds a registration card, the contractor must make a deduction from the labour element of any payments for the subcontractor's tax and national insurance liability.

Most 'contractors' and 'subcontractors' in the construction industry are affected by the scheme but these terms have a much wider meaning than normal. Consequently, government departments and agencies, local authorities, hospital trusts and other public bodies, airport operators and large businesses normally known as clients are 'contractors' under the CIS.

Private householders and businesses which spend less than £1 million a year on construction work are not contractors under the Scheme.

Proposals to change the Scheme were announced by the Chancellor of the Exchequer in the 2003 Budget.

References

Brook, M. (1993) *Estimating and Tendering for Construction Work*. Butterworth-Heinemann.

Canter, M. R. (1993) *Resource Management for Construction*. Macmillan Publications.

Cartlidge, D. (2002) *New Aspects of Quantity Surveying Practice*. Elsevier Butterworth-Heinemann.

CIB (1997) Construction Industry Board *Code of Practice for the Selection of Subcontractors*. Thomas Telford.

CIOB (1997) *Code of Estimating Practice*, 6th edn. Chartered Institute of Building.

Construction Task Force (1998) *Rethinking Construction*. Department of Trade and Industry.

Kwakye, A. A. (1997) *Construction Project Administration in Practice*. Pearson Education.

Okoroh, M. & Torrence, V. (1992) Knowledge-based Decision Support Systems in the Selection and Appointment of Subcontractors in Building Refurbishment. Technical paper based on PhD Thesis, University of Loughborough.

Sidwell, A. C. (1983) An evaluation of management contracting. *Construction Management and Economics*, Vol. 1, E. & F. Spon.

8 Planning for construction

8.1 Introduction

The tendering period is a busy time for the contractor and getting the price finalised and submitted on time is a high-pressure process. Once the tender has been delivered, there is a short period of anti-climax for the estimator in particular and the process starts again with the next enquiry.

The client team, on the other hand, is very much focused on the project in hand and they will be anxious to compare the tenders received with the budget and cost plan. If the tenders are too high, changes will have to be made to the design or specification in order to make savings.

When the preferred tender has been checked for mistakes or qualifications, the successful tenderer will be notified and contract documents prepared.

The time between contract award and taking possession of the site is frequently very short and clients often apply pressure to start 'yesterday'. Clients' advisors and contractors alike find this pressure hard to resist and often all concerned bend over backwards to get things moving.

This chapter concerns the contractor's processes in planning for the construction stage and in making sure that the contract proceeds in a timely and efficient fashion.

8.2 Planning procedures within a large organisation

The reasons for undertaking planning have been outlined in Chapter 4. During a major construction project, the contractor will need to implement appropriate procedures in order to keep the master programme under constant review. Figure 8.1 shows the relationships between the master programme, stage programmes and the 4–6 weekly and 1–2 weekly short-term programmes at site level.

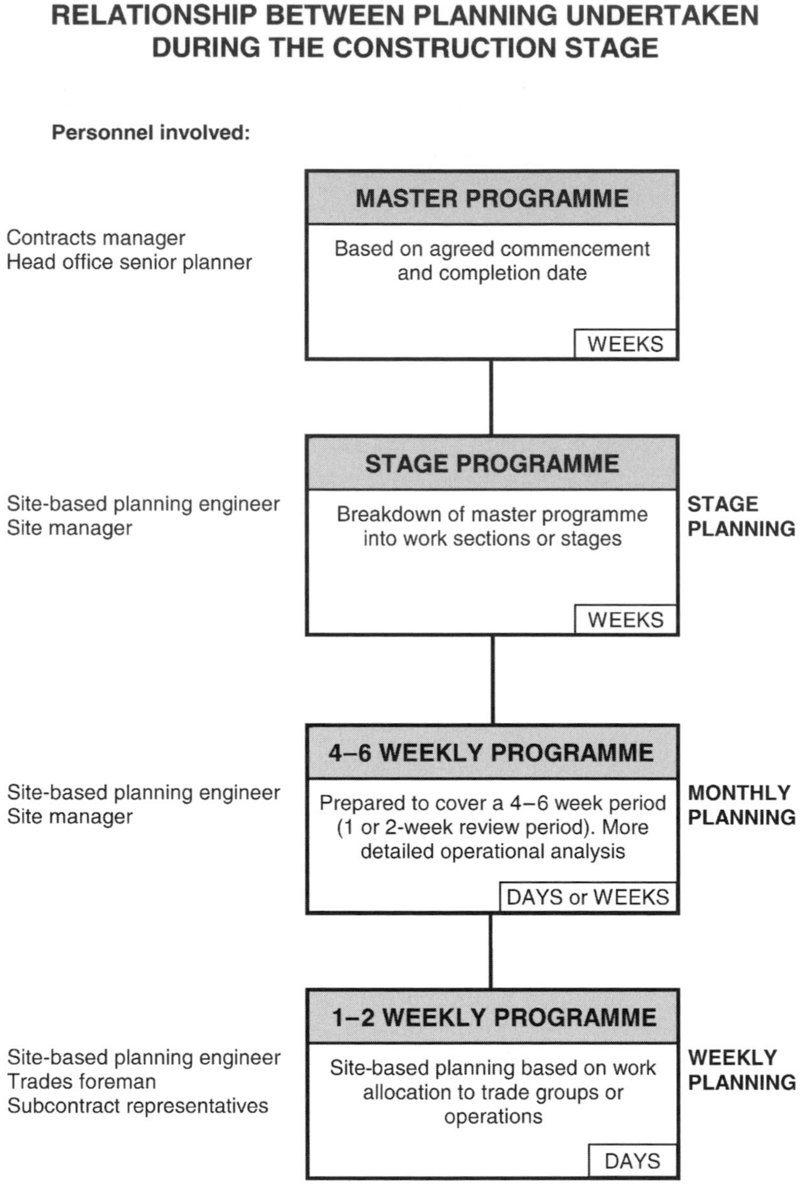

RELATIONSHIP BETWEEN PLANNING UNDERTAKEN
DURING THE CONSTRUCTION STAGE

Personnel involved:

Contracts manager
Head office senior planner

MASTER PROGRAMME
Based on agreed commencement and completion date
WEEKS

Site-based planning engineer
Site manager

STAGE PROGRAMME	
Breakdown of master programme into work sections or stages	**STAGE PLANNING**
WEEKS	

Site-based planning engineer
Site manager

4–6 WEEKLY PROGRAMME	
Prepared to cover a 4–6 week period (1 or 2-week review period). More detailed operational analysis	**MONTHLY PLANNING**
DAYS or WEEKS	

Site-based planning engineer
Trades foreman
Subcontract representatives

1–2 WEEKLY PROGRAMME	
Site-based planning based on work allocation to trade groups or operations	**WEEKLY PLANNING**
DAYS	

Figure 8.1

The organisation structure of the planning department within a large contracting firm is shown in Figure 8.2. This shows the relationship between the pre-tender planning and the contract planning functions. The contract planning staff may be site-based on the larger projects and on smaller projects may be serviced by planning staff from an adjacent major site.

ORGANISATION OF THE PLANNING FUNCTION WITHIN A LARGE CONTRACTING ORGANISATION

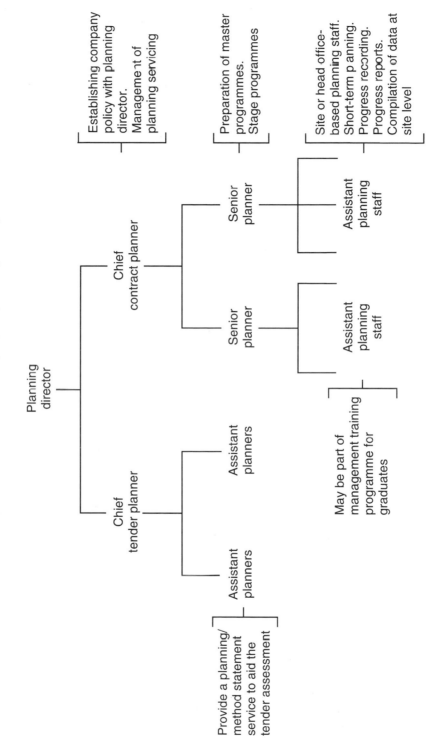

Planning director

Chief tender planner

Chief contract planner

Assistant planners

Assistant planners

Senior planner

Senior planner

Assistant planning staff

Assistant planning staff

Establishing company policy with planning director. Management of planning servicing

Preparation of master programmes. Stage programmes

Site or head office-based planning staff. Short-term planning. Progress recording. Progress reports. Compilation of data at site level

Provide a planning/method statement service to aid the tender assessment

May be part of management training programme for graduates

Figure 8.2

The contractor's planning procedures for the construction phase will now be discussed under the following headings:

- Pre-contract planning
- Contract planning

8.3 Pre-contract planning

On award of the contract the pre-contract planning process commences. The contractor may have up to 6 weeks in order to plan the commencement of works on site, or he may simply have 6 days. The commencement date will normally have to be negotiated with the client's representative or may be specified in the tender documents. On a major contract of, say, £3 000 000 in value, the following stages will be involved in the pre-contract planning:

- The pre-contract meeting and arrangements for commencing work
- Placing orders for subcontractors and suppliers
- Site layout planning
- Construction method statement
- Master programme preparation
- Preparation of requirement schedules
- Preparation of contract budgets
- Preparation and approval of the construction health and safety plan

It must be pointed out that no two companies undertake exactly the same procedures at the pre-contract stage. Procedures depend on the policy of the company and the need for establishing standard routines which may be followed on every new contract. The procedures outlined below could be considered normal in a larger company or in a more organised medium-sized organisation. Within the larger organisation a more formal approach will be taken to the arrangements for commencing work.

8.4 Pre-contract meetings

Prior to the commencement of the project, a series of pre-contract meetings will be held. For instance, a pre-contract meeting will be called by the client's representative in order to coordinate the start of works on site. It is important to establish lines of communication between the client's team and the contractor that are clear and transparent so as to avoid confusion and disagreement. Apart from obvious contact details, the contractor needs to know, for instance, who has authority under the contract to issue instructions, to certify payment and to agree the value of variations.

The pre-start meeting is set up to organise a multiplicity of matters including:

- Organisational and contractual details
- Dates of progress and valuation meetings
- Safety management arrangements
- The commencement of work on site

The contractor will also hold his own internal pre-start meetings in order to hand over the tender documentation and make arrangements for commencing work on site. It is also an opportunity to delegate responsibility to those entrusted with the duty of organising and managing the project.

The pre-contract arrangements outlined here relate to those undertaken by the contractor on a traditional contract and will require consideration of the following:

- The pre-contract planning meeting
- The registration of drawings and distribution of information
- Preparation of the construction health and safety plan (in compliance with the CDM Regulations)
- Arrangements for commencing work

The contractor's pre-contract meeting

A meeting will be held to announce the award of the contract and allocate responsibility to the construction team responsible for undertaking the work. All documentation produced at the tender stage will now become available to the construction section. The following personnel will normally attend the meeting which may be chaired by the chief estimator or the contracts manager.

Chief estimator

The chief estimator will probably act as meeting chairman in the first instance as he is fully aware of all decisions taken at the tender stage. He will be responsible for handing over all the estimating data to the contracts section including:

- Estimate summary and analysis
- Build-up of the all-in rates and net bill rates
- Summary of subcontractors' and suppliers' quotations
- Pre-tender method statements
- Preliminaries build-up
- Pre-tender safety assessment
- Tender adjudication report

Contracts manager

The contracts manager will be responsible for organising the commencement of work. At this stage it is not always possible for the construction manager to be appointed or available as he may be tied up on another contract. On the larger multi-million pound projects it would prove advantageous, however, to have him allocated to the project team at the pre-contract stage.

The contracts manager is therefore responsible for all pre-contract activities within the medium and smaller-sized organisation. He would be responsible for finalising the operational method statement and assisting with preparing the master programme.

Since 1995 it has been a statutory requirement to prepare a health and safety plan for the construction phase of most projects and the client must be satisfied that this plan complies with the Construction (Design and Management) Regulations prior to allowing the commencement of work on site.

The contracts manager's responsibilities will also include making arrangements for the transfer of key staff to the project or the recruitment of additional staff to manage the project. Arrangements will have to be made for the delivery of site accommodation and the mobilisation of plant and equipment for the initial site operations.

Company buyer

The buyer, or quantity surveyor in the medium-sized organisation, is responsible for placing orders for subcontractors and suppliers based on the information received at the tender stage. It may be the policy of the company to ask subcontractors and suppliers to requote for the work now that the contract has been secured. Smith (1995) makes an interesting reference to the practice of 'bid peddling' once the contractor has been awarded a contract. This practice is often frowned upon by the subcontractor fraternity and rightly so.

In many of the medium-sized construction firms, the contract quantity surveyor undertakes the responsibility for placing subcontract orders and preparing contracts. This practice enables the surveyor to become familiar with the project from the outset. The placing of subcontract orders will, however, continue throughout the project period. The scheduling of key subcontract dates will need tying in with the master programme and relating to the project requirement schedules.

Chief quantity surveyor

The chief quantity surveyor or managing surveyor is responsible for allocating surveying personnel to the project and arranging for the checking and signing of the contract. The surveyor will also be responsible for preparing the contract cumulative value forecast based on an analysis of the contract bills and master programme (see Chapter 6).

Office manager

It will be necessary to establish communication channels for the distribution of project information as it is received from the architect and this will involve establishing procedures for the circulation of drawings and correspondence. This is normally the job of the office manager, who will also be responsible for allocating office staff to deal with wages and material invoicing.

He will also be responsible for sending an F10 notice of appointment of the principal contractor to the local office of the Health and Safety Executive, for issuing any statutory building notices to the local authority and for making applications for hoarding licences, footpath crossings and the provision of temporary service connections.

The necessary insurance requirements for contract specific cover will also need organising.

The client's pre-contract meeting

The client's pre-contract start-up meeting is usually chaired by the architect or engineer, and the design team, client's representative and the contractor will be present in order to establish initial contact between the parties. This enables channels of communication to be set up for the issue and distribution of project information.

At the pre-start meeting the contractor may be asked to present his outline programme. This may indicate the requirements for key nominated subcontract dates in order that realistic information requirements may be assessed. Outstanding matters in relation to the contract commencement dates may be discussed, together with arrangements for the signing of the contract if this has not been done.

A typical agenda for such a meeting might contain some or all of the following items:

(1)   Introductions
(2)   Apologies for absence
(3)   Employer's organisation and delegated powers
(4)   Contractor's organisation
(5)   Tax exemption matters
(6)   Insurances
(7)   Notices
(8)   Commencement date
(9)   Safety management arrangements
(10)   Programme and method
(11)   Site boundaries and access
(12)   Setting out arrangements
(13)   Working hours
(14)   Contractor's tip and cleaning roads
(15)   Communications and correspondence
(16)   Progress meetings
(17)   Valuations and payment
(18)   Emergency procedures
(19)   Any other business
(20)   Date of next meeting

8.5 Placing orders for subcontractors and suppliers

As indicated above, it will be necessary to consider company policy with regard to procedures for placing orders for subcontractors and suppliers. The practice of offering work on the basis of a Dutch auction should not be encouraged. Chapter 7 deals with subcontractor selection and control criteria.

Within larger organisations procedures are implemented to award subcontracts on the basis of standard forms of subcontract such as the JCT Standard Form of Domestic Sub-Contract 2002. Nominated subcontracts are usually let on the JCT Nominated Subcontract Form.

Within small and medium-sized organisations, subcontracts may simply be awarded on the basis of a letter of appointment with no formal written contract entered into. Alternatively, some contractors have their own bespoke conditions which subcontractors should read very carefully.

8.6 Site layout planning

Site layout planning is an essential part of pre-contract planning. The contractor is often required to submit his proposals for approval by the client's representative prior to commencing work on the project. It is important to consider the allocation of preliminaries facilities allowed in the original estimate as this establishes the basis of the preliminaries budget.

Overspending on the preliminaries is a common problem on contracts and one must learn to work within the moneys allocated at the tender stage – no matter how inadequate.

Where a keen price has been put in for the work at the tender adjudication stage, there will no doubt have been some reductions in the site overhead allowances.

Site layout plan check-list

Location of offices and site accommodation

- On open sites the site accommodation should be located close to the entrance in order that vehicles and personnel may be readily observed entering the site. Under the CDM Regulations the principal contractor is responsible for preventing unauthorised access. Site notices instructing personnel entering the site to report to the office, should be displayed. Site compound areas should be adequately stoned up or surfaced and consideration should be given to the site parking of vehicles for staff and operatives. The site office area may be fenced and provided with secure gates
- The space available on site may be at a premium and on restricted sites, similar to the one shown in Figure 8.3, consideration may be given to locating the offices on a gantry over the pavement or alternatively renting some space adjacent to the site. The contractor may also consider locating his offices in the building being refurbished, if this ties in with the sequence of work. Consideration may also be given to moving the accommodation into completed parts of the building as it is being constructed. On restricted sites problems also relate to the storage of materials and the location of plant and equipment
- The location of offices and accommodation will also be influenced by the location of access roads and site services. Careful consideration is required with regard to site security. Arrangements should be made to have the site well illuminated at night. Powerful security lighting may be used for this purpose in order to deter theft and break-ins. On a site in a rough district of Manchester this was recommended by the police, but during the first week of the project the security lighting was stolen. Consideration will have to be given to the siting and location of the site signboards and any sample panel areas (e.g. facing brickwork)
- Figure 8.4 shows the proposed site plan for the construction of a three-storey office block. The reinforced concrete frame is externally

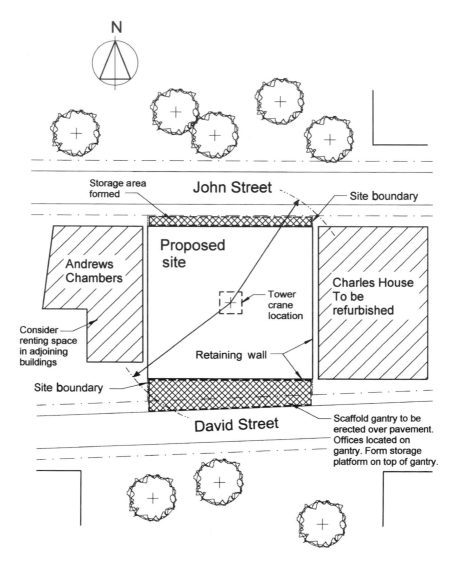

N

Storage area formed

John Street

Site boundary

Andrews Chambers

Proposed site

Charles House To be refurbished

Tower crane location

Consider renting space in adjoining buildings

Retaining wall

Site boundary

David Street

Scaffold gantry to be erected over pavement. Offices located on gantry. Form storage platform on top of gantry.

INNER CITY DEVELOPMENT

Figure 8.3

finished in quality brickwork. Figure 8.5 shows a proposed site layout plan for the project

Location of site services

- Consideration will have to be given to the location of existing and new site services in respect of water, drainage connections, electricity and telephone services serving and crossing the site

Description of construction method

Three-storey RC frame brick and block
external walls.
PC floors (wide slab).
Steel mansard roof structure.
Timber rafters/timber roof.
Ready-mixed concrete supply.
Timber stud walls internally.
Plasterboard finish.
Suspended ceilings.
Tower crane allocated for frame erection.

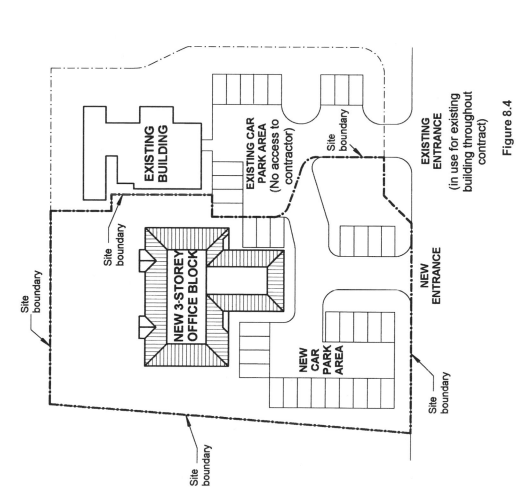

SITE PLAN
3-STOREY OFFICE BLOCK
(Not to scale)

Figure 8.4

ITEMS included on LAYOUT PLAN

Site security fencing.
Entrance gates.
A – Mess accommodation.
B – Tiered office accommodation.
C – Stores – lock-up.
D – Storage racking for finishing materials.
E – Brick storage area on hardstanding.
F – Formwork/reinforcement fabrication areas.
G – General hardstanding area for movement corridor – formed up on car park area at commencement of contract.
H – Area for subcontractors' accommodation and storage.
I – Site floodlight at high level.

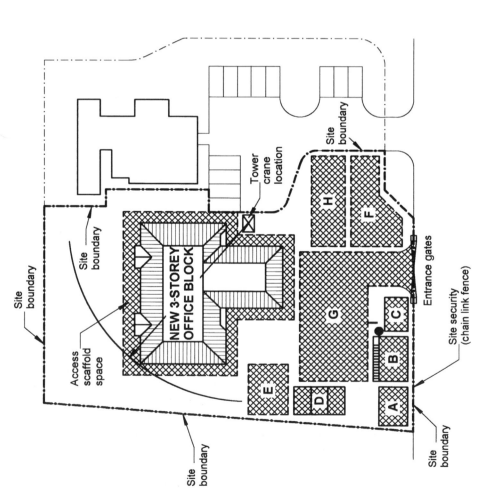

SITE LAYOUT PLAN
3-STOREY OFFICE BLOCK
(Not to scale)

Figure 8.5

- The statutory authorities will need to be contacted in order to establish existing locations and entry points for new services. The contractor will be required to establish the proposed location of temporary standpipes for site mortar mixing or facilities for the washing down of site vehicles. On open, greenfield sites, permanent services may have to be laid early in the project to provide temporary service connections
- Power supplies to major items of plant such as tower cranes will need consideration, together with the provision of electrical service connections for site power tools and site temporary lighting

Temporary roads, hardstandings and access

- Reference will again have to be made to moneys allocated in the contract preliminaries for these items. Temporary hardstandings may be required for the location of cranes and piling rigs or for the unloading of materials such as ready-mixed concrete
- On a factory project it may be necessary to provide a temporary access road around part of the building for mobile lifting or access equipment during the cladding stage of the works
- It may prove opportune to lay hardcore to car park areas early in the contract in order to use them as temporary hard standings for the works
- Foundations for tower crane bases or tracks will need preparing early in the contract prior to the erection of the cranes

Location of plant and equipment

- The contractor will have to consider the storage and security of plant during evenings and weekends. A locked compound may be provided. The location and working radius of cranes will need to be marked on the site layout plan in order to establish the best location for delivery points
- Consideration will have to be given to the rights of adjoining owners with regards to the swinging of tower crane jibs and booms over land adjacent to the site. The contractor has no rights to enter the air space of an adjacent site without permission from the owner, and may be sued in trespass if he does so. Likewise the contractor has a duty of care to persons using footpaths, streets and highways adjacent to the works. The main contractor is responsible for the deeds of his subcontractors when working in the vicinity of an adjacent highway. The contractor may be sued for negligence in the case of an accident involving the public
- Consideration will also have to be given to site space requirements around the building for the provision of scaffolding

Material storage areas on site

- One of the major considerations is the safe and secure storage of materials and components, in order to reduce waste and ensure safe working procedures
- Materials are an expensive commodity and care must be taken in their handling, storage, protection and placing in position. Considerations for material storage (lay-down) areas should be shown on the site layout plan
- Bricks and blocks should be stored on a firm clean surface or on pallets suitable for handling with a fork lift truck. Blocks should not be stacked too high. Bricks should be stored in locations which do not become waterlogged and where the brick stockpiles will not become contaminated by mud
- Structural timber should be ordered in cut or stock lengths to avoid cutting waste and may be banded for ease of handling. Timber must be stored clear of the ground and protected with polythene or tarpaulin sheets to avoid changes in moisture content
- Finishing timbers such as architraves, skirtings and door linings should be ordered in door sets. Skirtings and timber mouldings must be stored internally on horizontal racking. On housing projects and the like, garages should be constructed early in the project where possible and used for storage purposes
- Roof trusses may be stored on purpose-built, timber or steel storage racks, which support the trusses clear of the ground. It is preferable to deliver the trusses to site in phased deliveries so that they may be lifted directly into position
- Scaffolding and props may be stored on horizontal racking with the fittings stored in drums or bins. These are expensive components to lose as the contractor will be charged for losses by the hire firm
- Drainage goods are fragile and easily damaged if mishandled. Drainage fittings may be delivered to site in a crate. Pipes may be banded and stored on timber separators. Materials may be stored adjacent to the works in a small compound

8.7 Construction method statements

Construction method statements are prepared in order to explain the contractor's proposed working methods and demonstrate how the durations and sequence of work shown on the master programme will be achieved. At the pre-contract stage in the project, method statements will be prepared in detail in order to be sure that durations are accurate and reliable. Method statements may be presented as written statements or in tabular form.

The main uses for work method statements are:

- To calculate activity durations for the programme
- To decide on gang compositions and thus the resourcing requirements for individual activities
- To plan activities in detail so that a logical construction sequence is adopted
- To provide an easily understood document which can be communicated to those who will carry out the work on site.

Preparing method statements requires discipline and logic in that the steps needed to complete an activity have to be thought out. Hand in hand with this is consideration of what plant and labour will be used, as this is fundamental to achieving a sensible working method and sequence. Temporary works and working space requirements will also have to be considered, as will subcontractors' attendance requirements for packages such as piling and structural steel erection.

In order to calculate the duration of an activity, the output of the gang has to be considered. Outputs used by estimators are not suitable for planning as these relate to measured bill items where both main quantities and smaller items are included. In brickwork, for instance, the bills of quantities will include an item for facing brickwork, say, and there will also be items for laying the damp-proof course, for closing cavities and for building in lintels. All these items would make calculations too complex for planning purposes and thus 'incidental' items should be reflected in an overall output for the facing brickwork quantity alone.

The activity duration in days would then be given by:

$$\frac{\text{Quantity of brickwork in m}^2}{\text{Output of gang/hr} \times 8 \text{ hrs/day}}$$

Where output is expressed in m² per hour.

Alternatively, the quantity of bricks in thousands can be calculated and divided by the output of the gang in bricks laid per day. This is fairly straightforward but the difficult bit is to judge what the output should be. This will depend on factors such as:

- The gang make up
- The type of work
- Location of the work
- The standard expected in the specification

Figure 8.6 illustrates the plan and section of a reinforced concrete basement to be constructed as part of a multi-storey office development project on an open site. Figure 8.7 shows an operational method statement and Figure 8.8 illustrates a risk assessment statement for the initial stages of the work. The contractor will, of course, refer to any pre-tender method statements prepared.

Approval of construction methods may be a contractual requirement depending on the form of contract used prior to commencing any operations on site.

The construction method cannot be considered in isolation from safety, and part of the pre-contract planning process is to think about the provision of safe systems of work. This is a statutory requirement under the Health and Safety at Work etc. Act 1974. Safe systems are commonly expressed in the form of safety method statements.

Safety method statements covering the main stages of the works are prepared at the pre-contract stage. They will be based on risk assessments, which are a statutory requirement under the Management of Health and Safety at Work Regulations 1999. Whilst safety method statements are not required by statute, it is common practice to include them in the construction health and safety plan together with subcontractors' risk assessments and method statements. Certainly the client must be satisfied with the health and safety plan before allowing work to start.

There must be no deviation from a method statement as this can lead to confusion and possibly accidents. Where a change in working method is required, this should be discussed and the method statement changed formally before going ahead with the work. In this respect it is crucial that the workforce is notified either by a tool-box talk or task talk and any subcontractors notified of the impact on their operations. On no account must a method statement be changed on an ad hoc basis as there have been a number of multiple fatal accidents for this reason. One of these concerned the demolition of a multi-storey reinforced concrete building. The normal site foreman went on holiday and his stand-in came to site without being briefed. He adopted a method of working which was contrary to the agreed method and an 18 tonne excavator fell on to the floor below as a result. Two workmen were crushed to death.

Risk assessments and safety method statements may be combined and this principle is illustrated in Figure 8.8. Further examples are given in Chapter 10.

8.8 The master programme

The contract master programme is an important management tool. It is an essential requirement in the coordination and control of the many integrated tasks to be undertaken during the project and is also used by

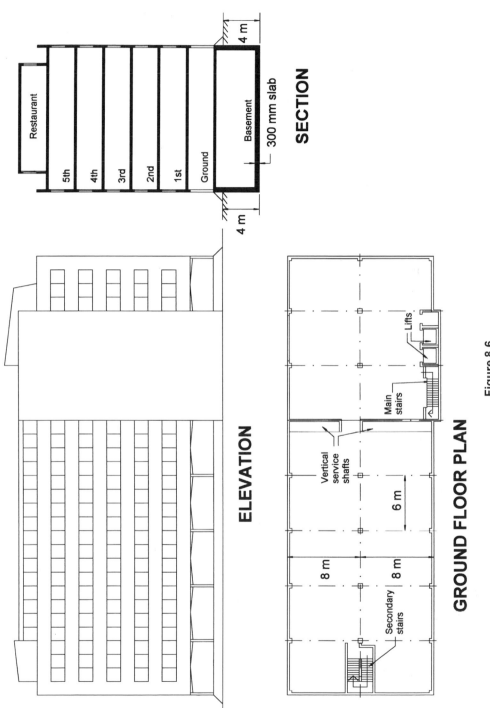

SECTION

300 mm slab

4 m

4 m

Restaurant

5th

4th

3rd

2nd

1st

Ground

Basement

ELEVATION

GROUND FLOOR PLAN

Lifts

Main stairs

Vertical service shafts

6 m

8 m

8 m

Secondary stairs

Figure 8.6

BASEMENT TO MULTI-STOREY BUILDING

CONSTRUCTION METHOD STATEMENT

OPERATIONAL STATEMENT

Page 1/2

OP No.	OPERATION	QUANTITY	METHOD	RESOURCES PLANT	RESOURCES LABOUR	NOTES
1	Reduced level excavation (0.5 m deep)	500 m³	Excavate to reduced level using hydraulic backactor (JCB 6C). Load 10 m³ wagons and remove to tip off site.	Hydraulic excavator 4 × 10 m³ lorries	1 - Banksman	Duration 2 days
2	Earth to support basement excavation	Free draining sand/gravel WT 1.5 m	Battered excavation. Wellpointing to perimeter of basement. 150 dia. header – 2 wellpoint pumps. 50 mm pvc risers – 80 to 100 points. Wellpoints to be in operation three days prior to commencement of excavation work.	Hired wellpoint equipment	3 - Labourers	Installation period 2 weeks (10 days)
3	Excavation to basement (including batter)	4800 m³	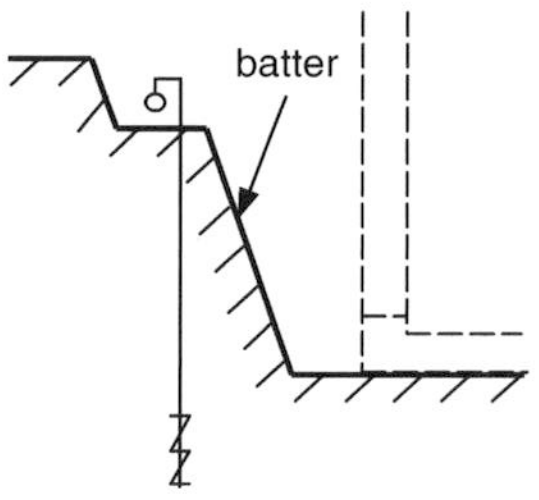 Excavate using large hydraulic backactor loading wagons. Part deposit on site, majority cart away to tip off site. batter	Hydraulic excavator 4 × 10 m³ wagons	1 - Banksman $\dfrac{4800}{40} = 120$ mhrs $\dfrac{120}{8} = 15$ days 15 days = 1 machine 8 days = 2 machines	Output 40 m³/hr

Figure 8.7a

BASEMENT TO MULTI-STOREY BUILDING		CONSTRUCTION METHOD STATEMENT	OPERATIONAL STATEMENT		Page 2/2
OP No.	OPERATION	QUANTITY	METHOD	RESOURCES		NOTES
				PLANT	LABOUR	
4	Basement floor (300 mm thick)	250 m³	**Formwork** Plywood panel shutters to edge of floor formed with 150 mm high upstand kicker.		2 - Joiners	
			Reinforcement Steel to be delivered to site cut and bent. Lift reinforcement into basement area using crane.	Low pivot jib crane	4 - Steel fixers	
	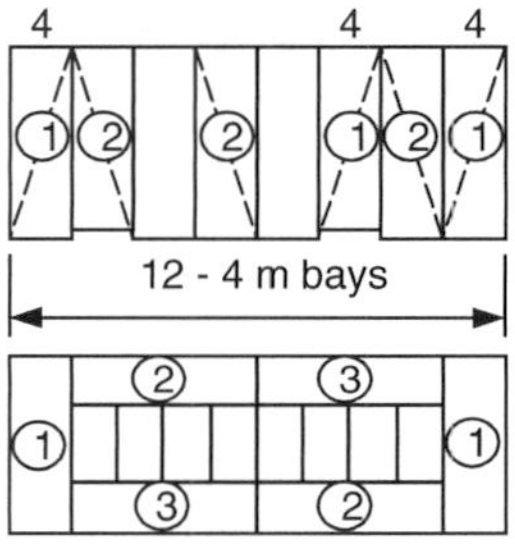		**Concrete** – Ready-mixed supply. Divide basement area into 4 m wide bays. Pour in sequence shown on bay layout. (Alternative proposal shown) Place concrete using concrete pump, pouring 3 bays per day (20 m³ per bay). End bays to be poured in first sequence in order to release work on wall reinforcement and formwork.	Concrete pump (hired) 2 Poker vibrators	6 - Labourers	Pour 3 floor bays per day

Bay layout

12 - 4 m bays

Alternative proposal

Figure 8.7b

| | | RISK ASSESSMENT STATEMENT | | Page 1/1 |

BASEMENT WORK

OP No.	OPERATION	CONSTRUCTION HAZARD LIKELY TO BE ENCOUNTERED	PERSONNEL AFFECTED	DEGREE OF RISK	PROPOSED CONTROLS TO REDUCE RISK
1	Reduced level excavation	Moving vehicles	Public/ own labour	Low	Traffic management system Segregate pedestrians from vehicles Provide adequate signs
		Mud on road	Public	Medium	Provide wheel washer Provide road sweeper
		Moving plant	Own labour	Medium	Banksman to control movement of plant and vehicles Provide adequate stop barriers at edge of excavation.
		Plant security	Children/ public	Low	Plant to be disabled during evening/night. Site security services provided.
2	Excavate to basement	Deep excavations	Subcontract labour Children/ public	High	Excavation to be staged Batter to be formed to correct slopes. Excavation to be adequately fenced off. Timber stop barrier around top of excavation. Provision of wellpoint system. Care in use of jetting and wellpoint installation system. Operatives to be trained by specialist plant installer.
		Moving plant during excavation works	Own labour Subcontract labour	Medium	Provision of banksman to direct lorries into position for loading by excavator. Provision of adequate hardstanding for plant and vehicles.
3	Concrete to basement floor	Wet concrete	Own labour	Medium	Provide suitable PPE: • wellington boots with steel toe caps and soles • high visibility jackets • heavy duty rubber gloves • hard hats Provide suitable and sufficient washing/ drying facilities in accordance with Construction (HSW) Regulations

Figure 8.8

the client's contract administrator to monitor the contractor's progress. On a major project, many different programmes are required to cover each stage of the work in order to ensure the smooth flow of information during the project. This includes consideration of:

- A contract master programme covering the major phases or sections of work and clearly indicating the planned sequence of construction
- A programme indicating the key dates for the release of design team information in order to meet the requirements of the contract master programme
- A programme to coordinate the requirements of subcontractors, material supplies and the resources of the main contractor
- Separate detailed programmes relating to the various project phases or stages highlighting the links between each work stage

The master programme forms the basis of the contractor's budgetary control and financial forecasting procedures and aids the client in assessing his cash funding requirements at the monthly payment stages. It is also important in relation to the contract.

Despite its importance, the master programme is not usually part of the contract documents. Under the JCT 98 Standard Form, for instance, these normally consist of:

- the articles of agreement
- the contract drawings
- the bills of quantities (or the specification in the Without Quantities version)

Including the master programme or a method statement in the contract documents is fraught with problems. This was discovered in the case of *Yorkshire Water Authority* v. *Sir Alfred McAlpine & Son (Northern) Ltd* (1985) in which the contractor was relieved of his obligations to carry out the works because they were physically impossible.

This is an entirely different situation, however, to a contractual obligation to provide copies of the master programme. For instance, under JCT 98 clause 5. 3.1.2, 'the contractor without charge is to provide the employer with two copies of the master programme for the execution of the works'. The contractor is also obliged, 'within 14 days of any decision by the architect in relation to an extension of time', to provide two copies of any amended programme. However, this clause only applies where the contractor has actually prepared a programme, which, unlike ICE Clause 14, he is not obliged to do.

If the master programme was included as a contract document it would somewhat impair its flexibility and usefulness as a management tool. It would involve both parties having to strictly adhere to the programme and the contractor would be obliged to start and finish each operation by the programmed dates and in the programmed sequence or risk being in breach of contract.

Consideration should also be given as to whether or not the master programme constitutes effective notice of the contractor's information requirements to the architect. In the case of *London Borough of Merton* v. *Stanley Hugh Leach* (1985) for instance, the judge held that the programme, if it gave sufficient detail, would constitute effective notice of the contractor's requirement for further information. It follows that a post-contract programme is a unilateral declaration of the intention by the contractor and notice of when he requires information provided that the notice is not unreasonably premature.

Notwithstanding contractual obligations, it is good practice for the contractor and design team to agree amicably a realistic contract programme prior to the commencement of the project. Copies of the master programme should be circulated to the architect for courtesy approval and also to the major subcontractors to indicate their approximate commencement dates and periods on site. This is particularly important when nominated subcontractors are involved and will avoid recourse to legal action which can only lead to strained relationships, which everybody can do without.

The master programme may be presented in bar chart, network or precedence format depending on the programming techniques adopted by the contractor. Contractors tend to use the programming technique which best suits their mode of operations, and works for them.

The programme should show the contractual possession and completion dates for the project, together with the main work activities on site. The master programme indicates the sequence of operations, and relationships may be shown between related operations by introducing links in the form of a linked bar chart.

The programme is used to record progress weekly and monthly throughout the contract period in order to achieve the planned completion date.

Figure 8.9 illustrates an extract from a master programme showing three stages of work on a steel framed building project:

- Site establishment = 7 weeks
- Substructure   = 9 weeks
- Superstructure   = 14 weeks

A single bar line distinguishes the overall duration of each stage. The programme also shows the subtasks within each main or summary activity. Using project management software, the subtasks can be 'rolled-up' to produce a simpler overview of the master programme.

MASTER PROGRAMME

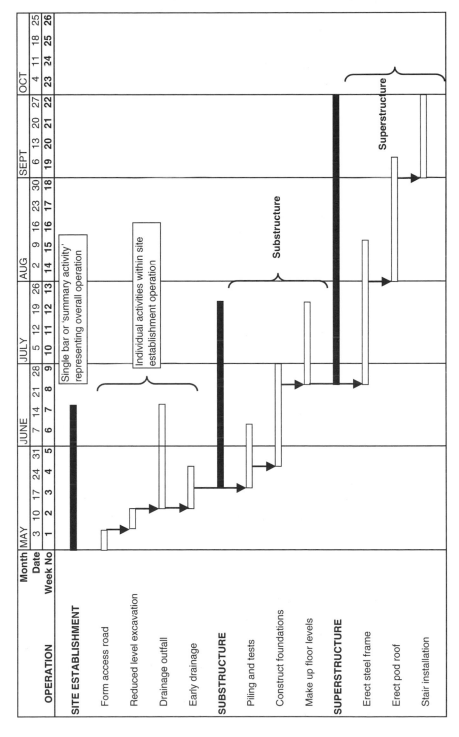

Figure 8.9

Finish to start and start to start links have been shown to link the various stages of work.

8.9 The target programme

During the pre-contract period, it is common practice for contractors to develop a target programme as well as a master programme. The target programme effectively compresses the master programme into a shorter time period, thereby saving on time-related costs and releasing resources sooner for the next contract. For instance, the master programme may show completion in 22 weeks but the target programme will indicate completion in 17 weeks, thereby saving 5 weeks (see Figure 8.10).

In order to achieve the target programme it must be realistic and not just hopeful. It therefore needs to be based on the informed opinion of the construction team, and especially the contracts manager or site agent, that the project can realistically be completed faster than the time shown on the master programme.

The implications are far reaching in terms of the contractor's own supervision and resources, the procurement and management of sub-contractors and the ability or willingness of the design team to provide the necessary drawings and other information in time.

Following the judgment in the case of *Glenlion Construction* v. *The Guinness Trust* (1987), a programme showing a shorter completion period than that required under the contract, even though it may be annotated with information deadlines, will not be successful in putting pressure on the architect or engineer to produce drawings etc. to a tight time-scale.

Any claim by the contractor based on the failure of the contract administrator to provide information, and thereby causing delay, will not be valid unless the contract is delayed beyond the *contractual* completion date. The architect or engineer is only obliged to furnish drawings and instructions 'within a reasonable time of the conclusion of the contract' and Keating (Furst & Ramsey 2000) takes the view that the employer is not under an implied obligation 'to enable the contractor to complete by the earlier date'. Therefore, 'provided that the contractor can still complete within the contract period, he cannot recover prolongation expenses' and 'the employer is under no obligation to pay compensation if the contractor is unable to achieve an accelerated programme'.

Target programmes place added pressure on the site management team and subcontractors alike and a careful balance needs to be struck between the demands of faster working and the health, safety and quality standards expected.

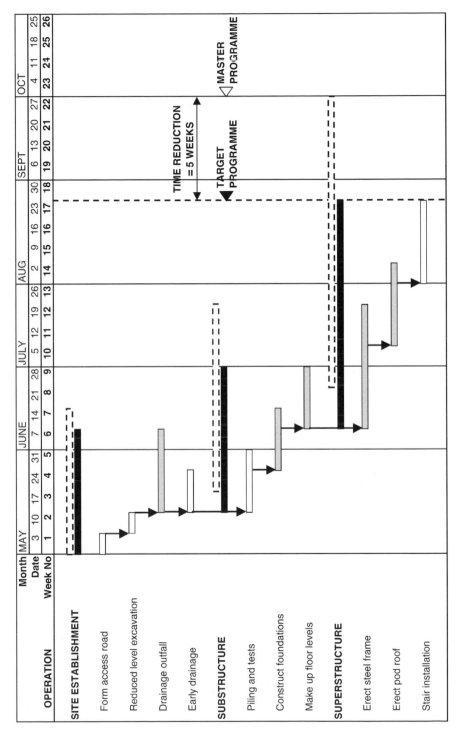

Figure 8.10

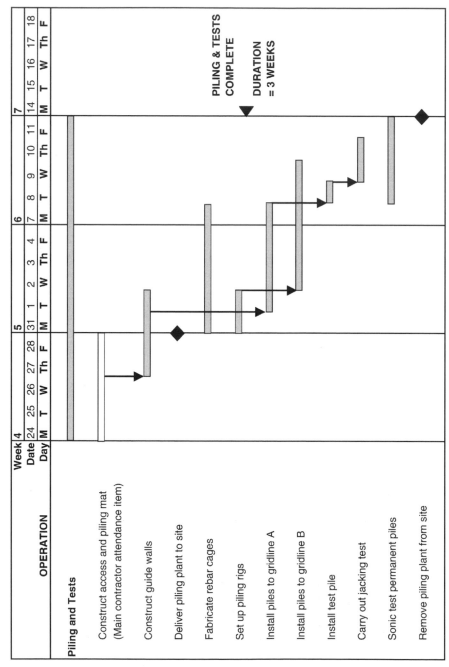

Figure 8.11

8.10 Subcontractor programmes

The master or target programme frequently shows subcontract packages as a single bar line simply because it is impossible to show every site activity in detail – the programme would be too complex and unwieldy. This principle is demonstrated in Figure 8.10 which shows, for example, 'piling and tests' as a single bar on a target programme with a duration of 3 weeks. However, each subcontract package is effectively a subproject with its own list of activities and events and the contractor needs to exercise control over them.

Contractors, therefore, commonly prepare subcontractor programmes for this purpose and an example is given in Figure 8.11. In this example, the piling and tests package is shown in detail with several activities and events happening within the overall duration of 3 weeks indicated on the target programme.

8.11 Procurement programmes

Figure 8.12 illustrates a procurement programme where the bar-lines represent the procurement periods for each major component supplier or subcontractor. Each bar line has been flagged with three milestone symbols relating respectively to information requirements, placing the order and the commencement of the operation on site.

The following examples illustrate how the procurement programme may be developed into specific programmes for the procurement of the 'steelwork' and 'finishes' packages.

Steelwork

Figure 8.13 shows a detailed bar chart display indicating the procurement requirements for the steelwork activity using a simple linked bar chart and early warning symbols. In practice the steelwork activity would be shown as a single bar on the main procurement bar chart. However, by using a project management software package, the detailed programme may be accessed simply by clicking on the bar line or, on some packages, by clicking on an icon to expand the activity and show the detail.

Finishes

Figure 8.14 shows a procurement bar chart for the finishes operation depicting lead-in times incorporated within the bar lines. Lead-in times are for the benefit of both the contractor and those providing the

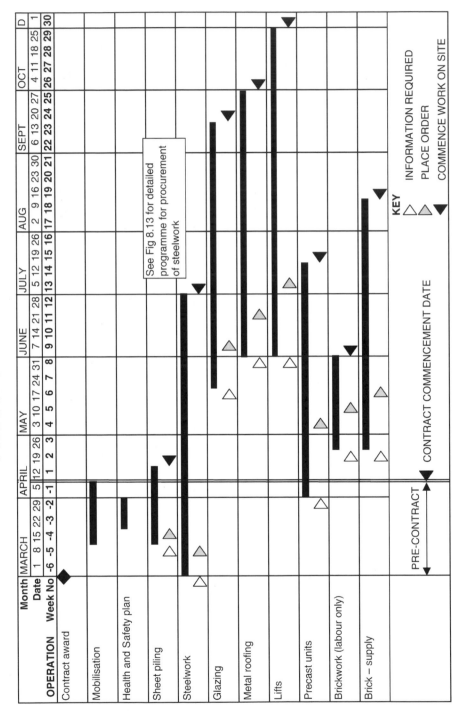

Figure 8.12

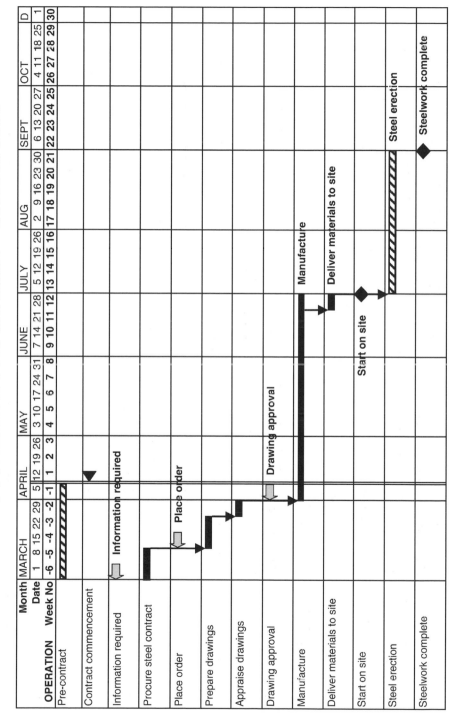

Figure 8.13

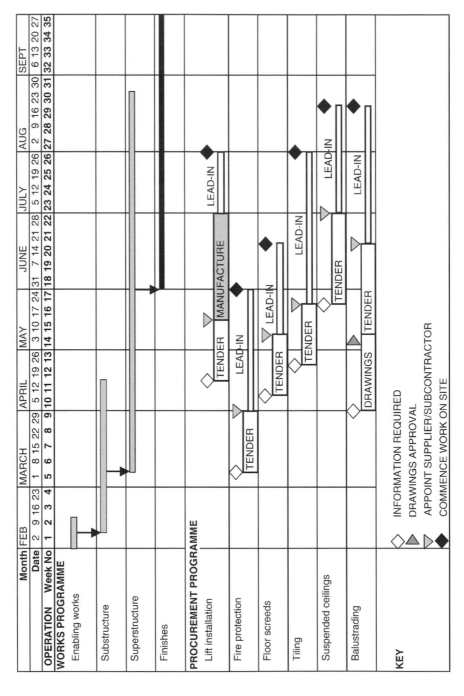

Figure 8.14

information. This might be the project architect or other designer or the design and build designer engaged by the contractor.

Each of the above examples clearly indicates 'must have' or 'must do' information reminders. It is no use producing key requirement dates if nobody keeps to them, but for this to happen dates must be realistic and not simply established to strengthen contractual claims. Management action must be taken when key dates indicated by the milestone symbols are not met or if the data is not available on the specified date. Delays resulting from the late release of information from whatever source must be confirmed in writing at the appropriate time in order to have contractual effect.

8.12 Requirement schedules

A long-standing bone of contention between contractors and contract administrators is the thorny issue of the provision of information and this has been the cause of many claims and disputes. The problem is especially prominent with conventional contracts where the design is incomplete at the tender stage.

The purpose of requirement schedules is to aid the smooth running of the contract by providing a programme for the preparation and release of information by the architect to the contractor throughout the project.

The importance of this is emphasised in the JCT 98 Standard and Intermediate Contracts which make specific provision for an Information Release Schedule. In JCT 98, the Articles of Agreement provide for an option (Recital 6) which binds the employer (client) to provide such a schedule saying what information the architect will release and when. This is also referred to in clause 5.4.1. Where the contract does not provide for a schedule, however, the onus is on the contractor to give notice to the architect sufficiently in advance of the date when the information is required (clause 5.4.2). Clause 5.4.2 also allows the architect to use his own judgement as to when information will be released, but no contractor worth his salt would be happy with this arrangement.

In any event, there is an important linkage between information release and claims. Under JCT 98 for instance, failure of the architect to comply with the Information Release Schedule provisions is a Relevant Event under clause 25.4.6 (extensions of time) and is included in the List of matters for claiming loss and expense (clause 26.2). Where there is no Information Release Schedule, there are nevertheless similar provisions for extensions of time and loss and expense where the architect is late giving instructions (clauses 25.4.6.2 and 26.2.1.2 respectively).

However, it is pointless asking for information relating to painting schedules at week 1 on a 70-week project and expecting to formulate a claim at week 20 simply because the information has not been received.

Many contractors may consider that 'the claims start here' as far as requirement schedules are concerned but this should not be the reason behind requesting information requirements at the pre-contract stage of the project. Surveyors, however, remain mesmerised by information requirement schedules!

Certain schedules relate to the contractor's internal requirements, while others concern key dates for the release of information from the client or client's representative. The following examples illustrate the types of requirement schedules in use.

Contractor's internal schedules

- Key materials schedule
- Plant schedule
- Domestic subcontractor requirements

Client-based information requirements:

- Subcontract schedule (nominated subcontractors and suppliers)
- Drawings and information release schedule
- Request for information sheets

Key materials schedule

Figure 8.15 illustrates the key materials schedule which is prepared in conjunction with the contractor's planning or buying section and shows the key materials scheduled to meet key operational programme dates. The material requirements are assessed from the contract drawings, with the quantities in the bills being used as a check.

Phased deliveries may be arranged for materials such as bricks and precast floor units. On sites with restricted space, or inner city refurbishment projects, just-in-time deliveries to suit the contract programme may be preferred.

It is important that the material schedule contains full information with regard to what is expected of the supplier. Additionally, site management personnel must be encouraged to use and update the schedules as it is important that material suppliers are kept informed of the progress position with regard to any amendments to planned delivery dates.

MATERIALS SCHEDULE

							SHEET No.:	Date issued:	
CONTRACT:				PREPARED BY:					
CONTRACT No.:				PROGRAMME REF:				SUPPLIER DETAILS	
Material	B of Q ref.	Latest req. date	Latest order date	Delivery period	Required on site date		Supplier	Tel. no.	Order no.	Comments/ remarks

Figure 8.15

Plant schedule

The plant schedule indicates key dates for the major items of plant allocated to the contract (see Figure 8.16). The schedule may be presented in tabular format or as a bar chart programme. The schedule enables the plant department to plan its resources between the various projects operating at any one time. It also enables the right balance to be struck between using contractor-owned plant and hired-in plant from specialist firms.

A careful check must be kept on plant expenditure compared to the allowances for plant within the contract preliminaries. Overspending on plant is common at site level mainly because site managers can always find an excuse to keep plant on-hire for a bit longer than intended.

Subcontract schedule

The coordination of subcontractors and suppliers is an essential part of project control. The schedule illustrated in Figure 8.17 enables the site manager to overview key contract dates with respect to progress. Details are also shown relating to the subcontract order, contact address and notification dates. Reference should be made to the subcontractor's programme prepared at the contract stage. The late release of key information relative to nominated subcontractors and suppliers may lead to an extension of time when the delay is caused by the architect.

Figure 8.18 illustrates an information requirements schedule requesting information in respect of work packages. The responsibility for providing the information on behalf of the client's design team is indicated on the schedule.

Drawings and information requirement schedule

It is essential to monitor and record the receipt of contract drawings and other information issued by the architect. It is the contractor's responsibility to give adequate notice of information requirements such as:

● Setting out dimensions and measurements to site boundaries
● Reinforcement details for pile caps, foundation beams, etc.
● Details of ground floor services and pockets or fixing bolts
● Fixing details for cladding panels
● Door, window and ironmongery schedules
● Colour schedules for internal decoration
● Service layout details

PLANT SCHEDULE

| | | CONTRACT: | | | PREPARED BY: | | | | SHEET No: | | |
| | | CONTRACT No.: | | | PROGRAMME REF.: | | | | DATE PREPARED: | | |

Ref.	Operation	Plant item	Date requested	Date required on site	Duration on site	Plant release date	Actual release date	Notes/comments

Figure 8.16

SUB-CONTRACT SCHEDULE

CONTRACT:	PREPARED BY:				SHEET No.:	Date Prepared:			
CONTRACT No.:	PROGRAMME REF.:					SUBCONTRACTOR DETAILS			
Trade	Nom. or dom.	Latest inquiry date	Latest order date	Period of S/C notice	Comm. date on site	Subcontractor	Tel. no.	Order no.	Notes

Figure 8.17

INFORMATION REQUIREMENTS SCHEDULE

DETAILS REQUIRED	Contract Newcastle							Contract No		
	Design team							Workpackage requisition		
Work element/package	Arch	Str. eng	Serv. eng	Date required	Lead in period	Start on site		Requisition required by	Last date for order	Start on site
Demolitions	✔	✔		15 Dec	3 weeks	19 Jan				19 Jan
Foundations and lift pit	✔	✔	✔	5 Jan	4 weeks	2 Feb				2 Feb
Precast units	✔	✔	✔							
Raised access floor	✔		✔							
Suspended ceilings	✔		✔							

Figure 8.18

DRAWING AND INFORMATION REQUIREMENTS SCHEDULE

| | | CONTRACT: | PREPARED BY: | SHEET No.: |
| | | CONTRACT No.: | PROGRAMME REF.: | DATE PREPARED: |

Ref.	Information request	Date requested	Date required	Date received on site	Receipt of final details	Comments/remarks

Figure 8.19

When the architect fails to release information on time, the contractor may be entitled to an extension of time and loss and expense. The schedule shown in Figure 8.19 allows the contractor to record the date of the information request and compare this with the information release date.

Key information requirements may also be highlighted in the form of milestone events on the master programme. Alternatively the early warning system developed by a major contractor in the early 1970s is still a valid method of highlighting project requirements today. This is discussed in Chapter 9.

Request for information sheets

Figure 8.20 shows a written request for information of a more specific and detailed nature, i.e. drainage layout details relating to holes, pockets and ducts prior to pouring a ground floor slab area. This format allows the information requested to be matched with the information received on the same request form.

8.13 The construction phase health and safety plan

Before allowing construction work to start on site, the client must make sure that the principal contractor has a health and safety plan in place. Under the requirements of Regulation 15(4) of the CDM Regulations, the client has a statutory obligation to ensure that construction work is not started before the preparation of a health and safety plan which complies with the Regulations.

This requirement is subject to the test of reasonable practicability. In most cases it is probable that the planning supervisor will advise the client as to the suitability of the plan and have meetings with the principal contractor to discuss his health and safety management system for the project. Copies of risk assessments and method statements will invariably be requested.

The construction health and safety plan is the foundation upon which the health and safety management of the construction phase is based and the CDM Regulations Approved Code of Practice (HSE 2001) explains that it *'builds on information contained in the pre-tender plan'* and is developed from its pre-tender form by the principal contractor. The plan must basically consist of the principal contractor's arrangements to ensure:

- the health, safety and welfare of persons at work
- the health and safety of those affected by persons at work
- that construction risks are accounted for

REQUEST FOR INFORMATION SHEET

Contract	Wigan	Request date	15 Jan
Contract No	84/1	Prepared by	AB

Issued by	Issued to	Distribution
C. Wooton (Site Agent)	Architect – DBS Associates	Quantity surveyor Contracts manager File

INFORMATION REQUIRED	INFORMATION SUPPLIED
Drainage layout for ground floor slab to include: 1. Gully positions 2. Main service ducts in laboratory unit 3. Connections to toilets 4. Floor duct for waste water from machine hall	
CONTRACTOR'S COMMENTS	

Information required by	6 Feb
Date of response	
Further action date	

Signed	Signed
Date	Date

Figure 8.20

It is important that the client allows sufficient time between tender submission and starting on site for the principal contractor to fully develop his proposals for a health and safety management system for the project. The principal contractor's management system will include:

- Management organisation structure
- Communications
- Monitoring procedures
- Welfare arrangements
- Site rules
- Emergency procedures

- Site induction procedures
- Safe working procedures for early construction activities

With regard to specific operations on the contractor's programme, however, it is probable that design information may not be available at the start of the project and it may be that the health and safety plan cannot be fully complete before work starts. In such circumstances, therefore, it may be that only preliminary activities covering the first few weeks can be planned in any detail with appropriate risk assessments and method statements. Examples of such activities may include site clearance, demolition, groundworks, etc.

Later on in the project, detailed risk assessments and method statements can be developed for specific activities and included in the health and safety plan when sufficient design information is available or when specialist subcontract packages have been awarded. The CDM Approved Code of Practice recognises this possibility and that the health and safety plan will be developed over time as the job proceeds.

When the health and safety plan has been agreed, it is good practice for the client to sign his approval so that the principal contractor can confidently commence work on site without being in fear of breaching statute. Some contractors confirm this in writing anyway.

The role of the planning supervisor effectively ends when the design phase is complete but in practice this phase will overlap considerably with the construction phase. For instance, there will undoubtedly be design changes with a health and safety significance which the planning supervisor will have to coordinate, and there may be temporary works designs from the contractor which will require input from the planning supervisor.

On design and build projects, the contractor normally takes over the planning supervisor role from the person appointed by the client because the contractor assumes responsibility for the design and this will clearly be ongoing during the construction phase.

Consequently, the construction phase health and safety plan will be continuously evolving and the site manager will have to ensure that risk assessments and, where appropriate, method statements are prepared. This will include subcontractors' activities, especially where there is a specialist design input. As a result, site safety inductions will have to be updated and regular tool-box talks will be needed to communicate and discuss the plan with the workforce.

8.14 Contract planning

Once the contractor has commenced work on site, planning is required at regular intervals to determine when and how particular site activities are to be carried out. This type of planning is called contract planning and is carried out monthly or weekly (short-term planning).

Contract planning involves monitoring the master programme and updating it 'as built', reporting progress to management and making sure that the health and safety plan and all safety method statements for specific activities are up to date. Planning at this stage of the project is carried out at a fairly detailed level in order to flesh out the master programme and provide a basis for the detailed day-to-day arrangement of work on site. This helps to prevent inefficient working or duplication of effort on site and enables the contractor to make the best use of his resources.

During the contract stage, the following planning procedures may be followed:

- Stage programme
- Short-term programme
- Sequence studies
- Tool-box talks
- Task talks

Stage programme

The contractor's master programme is frequently developed as the project progresses in order to show a finer level of detail. Figure 8.21 shows a stage programme for part of the substructure activity shown on the master programme above.

The activity comprises three subactivities: piling and tests, construct foundations and make up floor levels. The construct foundations activity is shown in expanded form with logic arrows indicating dependency. Some activities (e.g. Fix rebar) are continuous and others (e.g. Blind bases) are intermittent.

Where alternative construction methods are being considered, the contractor might ask his quantity surveyor to prepare a cost comparison of the alternatives so that cost can be compared to the time allowed on the programme. There may be time benefits from choosing the more expensive alternative thereby saving on preliminaries and possible liquidated and ascertained damages.

Short-term programme

When the contract stage is reached, the level of planning becomes more detailed and many contractors prepare short-term programmes on a monthly or weekly basis. For example, a programme covering, say,

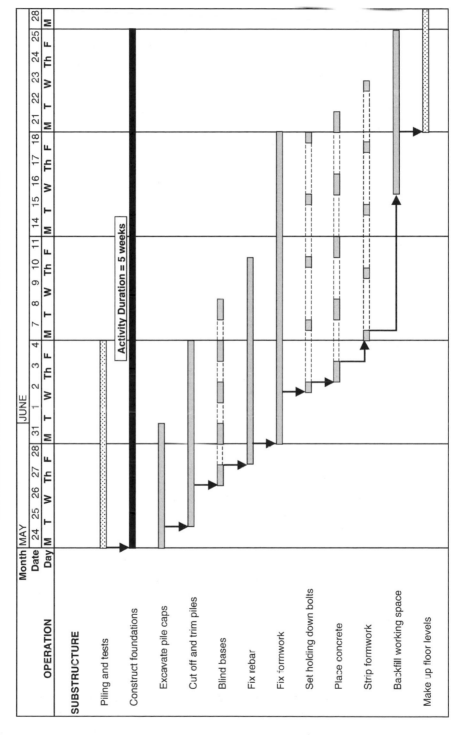

STAGE PROGRAMME – FOUNDATIONS

Figure 8.21

SHORT-TERM PROGRAMME

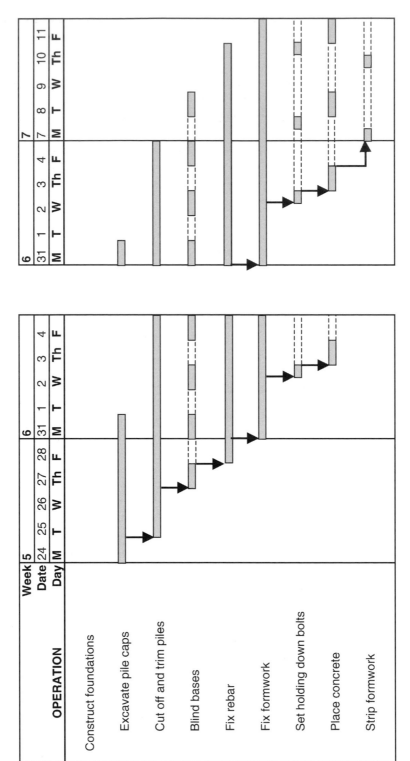

Figure 8.22

2 weeks' work might be produced and, at the end of the first week of this programme, another 2 week programme would be prepared for the following fortnight reflecting progress, problems and any changes. On refurbishment-type projects of a relative short duration, planning schedules may be prepared on a daily basis for each trade gang employed on the project.

Short-term programmes afford the contractor a much better means of controlling day-to-day operations on site and act as a useful method of communication between the site manager and the foreman, gangers or work package contractors. They can also be used in tool-box talks in order to discuss the health and safety implications of the programme with the workforce.

The principles of short-term planning based on a 2-week review period are illustrated in Figure 8.22 which shows the construct foundations activity from the stage programme. The programme for weeks 5 and 6 is prepared at the start of the project, and at the end of each weekly period, the short-term programme for the next 2-weekly period is prepared (weeks 6 and 7).

The short-term programme may be prepared by the site-based planning engineer or by the construction manager. Coordination meetings are usually held on a Friday afternoon with trades foremen and subcontractors' representatives and the work for the forthcoming 2-weekly period can be reviewed and the programmes prepared.

Responsibility for short-term planning may also be delegated to a site engineer when he is in charge of external works operations or specific concrete pours. This may be incorporated in the management training programme developed by the company.

Objectives of short-term planning

- To assist the coordination of operations in the short term, especially when considering the continuity of work for trade gangs and subcontractors
- To keep the master contract programme under constant review in the short term
- To highlight information requirements in the short term in order to meet planned completion dates for each stage of the work
- To assess key material requirements in the short term
- To keep senior site management informed of the progress position in the short term

8.15 Sequence studies

An important aspect of planning which concerns contractors is the efficient use of resources and nothing eats away profits more quickly than plant and labour working inefficiently or standing idle. Ensuring continuity of work is essential irrespective of whether the main contractor is doing the work or it is sublet.

Subcontractors are just as keen as main contractors to ensure continuity of work as they normally want to get in and out as quickly as they can. Good subcontractors are usually busy and want to move on to their next job quickly. Additionally, should the main contractor cause them delay or disruption or not provide attendances such as cranage and scaffolding on time, contractual claims will undoubtedly arise and this will have a damaging effect on the main contractor's bottom line profits.

All this is especially true where repetitive work is being carried out, such as housing or multi-storey construction. Following trades or operations cannot afford to be delayed by slower preceding activities, and attention has to be paid to this in the detailed planning of the project.

While perfect continuity of work is not possible in practice, sequence studies enable the contractor to analyse construction activities at the resource level in order to make best use of the labour and plant available.

The use of sequence studies is planning at a fine level of detail. However, where alternative methods of construction are being considered for a particular operation, a sequence study provides a means of determining the time and cost implications of the choices available.

A worked example of a sequence study for construction of the reinforced concrete basement for the City Road project is given in Case Study No. 1 in Chapter 10.

8.16 Tool-box talks

Tool-box talks are part of the contractor's safety management system and take place in the site cabin, usually between the supervisor or foreman and the operatives.

A tool-box talk affords an opportunity to raise awareness and promote a two-way discussion about day-to-day health and safety matters such as:

- General site health and safety
- Traffic management
- Roofwork and ladders
- Manual handling
- Noise and dust
- Abrasive wheels

- Cartridge-operated tools
- Buried services
- Asbestos

8.17 Task talks

Task talks are similar in principle to tool-box talks but they deal with specific site operations rather than general matters.

Typical task talk topics would include proposed construction methods for:

- Deep excavations
- The use and handling of drag-boxes for deep drainage
- Installation of sheet-piled cofferdams
- Hoisting and fixing external precast concrete cladding panels
- Demolition work by hand
- Fixing roof sheeting from a leading edge

The task talk gives the operatives who will be doing the work the chance to listen to the contractor's method statement and make suggestions for improvements based on their own previous experience.

References

Furst, S. & Ramsey, V. (2000) *Keating on Building Contracts*, 7th edn. Sweet and Maxwell.

HSE (2001) *Managing Construction for Health and Safety*, Approved Code of Practice and Guidance. HSE Books.

Smith, A. J. (1995) *Estimating, Tendering and Bidding for Construction*. Macmillan.

9 Project control procedures

9.1 Introduction

Like any other business, builders, contractors and developers have to plan and organise their day-to-day activities in order to manage effectively. Unfortunately, it is a fact of life that the best laid plans often go wrong and, when they do, they often go wrong in the worst possible way. This is called Murphy's Law.

Managers who can anticipate a problem before it becomes a crisis have more chance of making a success of their business than someone who takes the Mr Macawber approach of simply looking at excess income over expenditure as a measure of success.

Effective management requires control but all contractors have different ideas on the degree of control necessary for the projects they undertake. Many factors need to be considered, including the size and organisation of the firm and the scale and complexity of the projects in hand. However, in order to monitor performance, information needs to be collected within a structured reporting system, however simple, so that appropriate action can be taken if and when things start going wrong.

There are many aspects to the control of a business but, in terms of construction projects, three essential areas stand out for special consideration. These are the control of:

- Time
- Money
- Resources

This chapter considers each of these control areas in turn but first there is one feature common to all of them: the importance of keeping records.

9.2 Importance of site records

The keeping of site records during a project is imperative if a contractor is to receive his true entitlement as prescribed by the conditions of contract and also ensure that subcontractors are not successful with spurious claims for extra payment.

Records are necessary for:

- Establishing the basis of a claim for loss and expense
- Defending a counterclaim from a client
- Substantiating applications for extensions of time
- Claiming payment for extra work not included in the contract
- Reporting to progress meetings

Thomas (2001) lists 24 common types of records that contractors might keep, including site diaries, master programme updates and records of resources on site.

Records normally assume greater importance after the event than during it, and contractors often fail to keep adequate records during a project unless some formal procedures are adopted and maintained. On rare occasions, it has been known for a contractor to fabricate site diaries and other records if he is in a tight corner, rather than admit that he has no records at all.

Responsibility for keeping records on site

The contractor should delegate responsibility to his site management team for keeping, circulating and maintaining records and this includes site supervisors, planning engineers and quantity surveyors.

Construction supervisory staff should be responsible for:

- Maintaining a daily site diary
- Recording time lost per day due to exceptional rain, cold or heat. The time lost per week should be included in the report to the clerk of works, and minuted at monthly site meetings with the architect. This information may be used to establish a claim for an extension of time (JCT 1998, clause 25.4.2)
- Keeping records of the contractor's own labour and that of his subcontractors. Details relating to site labour are often noted in the daily site diary. Site labour records may be used as a basis of a counterclaim against a subcontractor due to his non-performance
- Recording verbal instructions from the architect or his clerk of works. A system of CVIs (confirmation of verbal instructions) is normally implemented by the contractor with copies sent to the contract surveyor
- Records of additional excavation works, especially where underground obstructions have been encountered. The effect of delays in foundation works may lead to the issue of a variation order and

hence to a possible extension of time (JCT 1998, clause 25.4.5). Photographic records may easily be recorded on a digital camera in order to record site conditions during foundation works. Excavation depths should be checked with the clerk of works as work proceeds, and a signature obtained

- Records are to be kept in relation to the issue of drawings from the architect or the consulting engineer. Procedures should be set up in relation to the circulation of drawings to subcontractors and other specialists. Responsibility for the keeping of a drawing register should be allocated to one of the construction team
- Materials received each day or week should be recorded on goods received sheets and copies sent to head office
- Daywork records (often referred to as the 'Daily Liar') are the responsibility of construction management. These may be used by the contract surveyor to establish fair rates

Site planning personnel should be responsible for:

- Updating short-term programmes each week or fortnight
- Recording progress each week on all the relevant contract programmes, i.e. weekly and monthly site master programme and subcontractor programmes. Preparing weekly and monthly progress reports for submission to the contractor's project manager
- Preparing records for the contract 'as-built programme' as work on site progresses
- Recording the effect of programme delays on the final contract completion date

Site-based surveying staff are responsible for:

- Keeping records of all site measurements which relate to the final account, i.e. records of foundations and drainage remeasurement. It may prove valuable to photograph complex drainage connections for reference purposes
- Recording details and measurement of individual variations to contract. Recording any delays and disruption resulting from undertaking work in variations. This may be linked to the construction planner's records
- The pricing of site daywork sheets and obtaining the appropriate signatures
- Correlating instructions received from the architect or his representatives with the CVIs issued
- Undertaking the reconciliation of the bills of quantities items with quantities abstracted from the site working drawings, e.g. brickwork and blockwork quantities

9.3 The control of time

The control of time will be discussed under the following headings:

- The role of the programme
- Progress recording
- Delay and disruption
- Extensions of time

The role of the programme

Planning is one of the functions of management and a programme is a key tool in the management process. However, very few plans turn out as expected and this is especially the case in construction work. Unforeseen ground conditions, delays due to weather and waiting for instructions are just some of the many reasons why this may happen.

It is therefore crucial to ensure that the planning process includes an element of control so that the plan can be monitored in the light of prevailing circumstances. Additionally, activities may have to be re-arranged or additional resources may be required to cope with difficulties encountered and a forecast of their likely impact may also be required. This is the role of the contract master programme which both the contractor and the client's contract administrator can use to monitor what is happening.

The contractor will normally have a contractual duty to complete the works on or before the date for completion stated in the contract and if he fails to achieve this without a valid reason he may suffer the deduction of liquidated and ascertained damages from his contract payments. This is of little benefit to the client, however, and therefore the contract administrator will use the master programme to check that the contractor is progressing satisfactorily. If things are not going according to plan, the architect or engineer will point this out at the monthly meeting and ask the contractor what he is going to do about it.

Failure to make satisfactory progress could lead to the replacement of the contractor and termination of his employment under the contract.

9.4 Progress recording

It is of little benefit producing a bar-chart programme, putting it up on the site cabin wall, and all sitting around looking at it. For the bar chart to be an effective management tool it is necessary to record progress on it and critically analyse the operations which are ahead or behind programme. The self-progressing bar chart has not yet been invented!

The purpose of progress recording is to provide a record of weekly and monthly progress in order to assist the construction director, contracts manager and site manager.

Procedure for recording progress

As a minimum requirement the planning engineer or site manager will monitor the progress on site on a weekly or monthly basis and produce a monthly report. The report should be completed 2 days before the site project review meeting and should consist of a visual progress report generated from the programme.

The report summary should highlight operations of concern, i.e. those which are out of sequence or behind programme. Comments should be added relating to the action to be taken in order to get the contract back on programme. Reasons for the delays may also be indicated, especially where the delays have been caused by variations to the contract, late receipt of information, or delays by specific subcontractors.

The monthly progress report should be issued to the contracts director, chief planner, contracts manager, site manager, site-based quantity surveyor and the design and build coordinator where applicable.

Progress recording procedures in practice

In some medium and large contracting organisations, laptop computers have been issued to senior site staff. The site manager will have access to the contract master programme and may have attended in-company training programmes in order to learn how to progress record operations on site. Thus the manager is able to progress record the master programme, operation by operation, and e-mail the information to his contracts manager for a mid-Monday morning progress meeting in the head office. This is an innovative approach to getting the site manager involved in IT procedures.

Progress recording on the bar chart

There are various ways of recording progress on the bar chart at weekly or monthly intervals. Responsibility may be delegated to the site-based planning engineer on the larger project, or the site manager on the smaller project.

As an additional safeguard the planning department at head office may act as an independent monitor of progress, undertaking regular visits to site and preparing a separate report on the progress position. This procedure forms part of the company's overall site management reporting procedures. It also alleviates the problem of site management personnel bending the truth a little when it comes to reporting progress.

Progress recording by colour coding

It may be policy within the company to record progress by colour coding on the bar chart. For example, green could represent the actual time expended on the operation and red could be used to indicate the percentage of the operation completed.

An alternative method of recording progress is illustrated in Figure 9.1. This is based on allocating a different colour to each week of the project and shading the work undertaken in that particular week with that colour. In this way the progress can be readily observed on the bar chart. This method may form the basis of an as-built programme.

In most contracts, progress is expressed as a percentage, on an operation-by-operation basis. The bar chart may also show the planned and actual percentage complete at each reporting date.

Progress recording by computer

Many software packages allow the user to progress record by computer. This is usually done on the basis of the percentage completed for each operation, which involves the planner simply entering the percentages, pressing the analysis command and leaving the rest to the computer. Figure 9.2 illustrates a bar chart progress analysis based on a widely used project management software package.

Progress recording by earned value

The physical progress on a contract can also be measured according to the earned value of work carried out, using a value envelope. This is established by plotting the forecast value for the contract based on both the earliest and latest times that work activities can be started. The boundaries of the envelope delineate the intended programme in money terms. The principles of this were explained in Chapter 4.

The monthly valuation is plotted on the envelope and, where the earned value is within the envelope, the contract is progressing according to the programme. Should the earned value fall outside the envelope, this signifies that the contract is either ahead or behind programme.

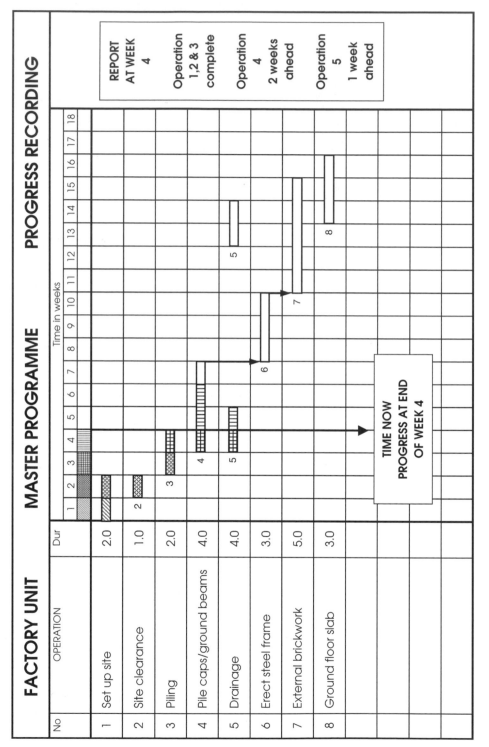

Figure 9.1

PROGRESS RECORDING BY COMPUTER

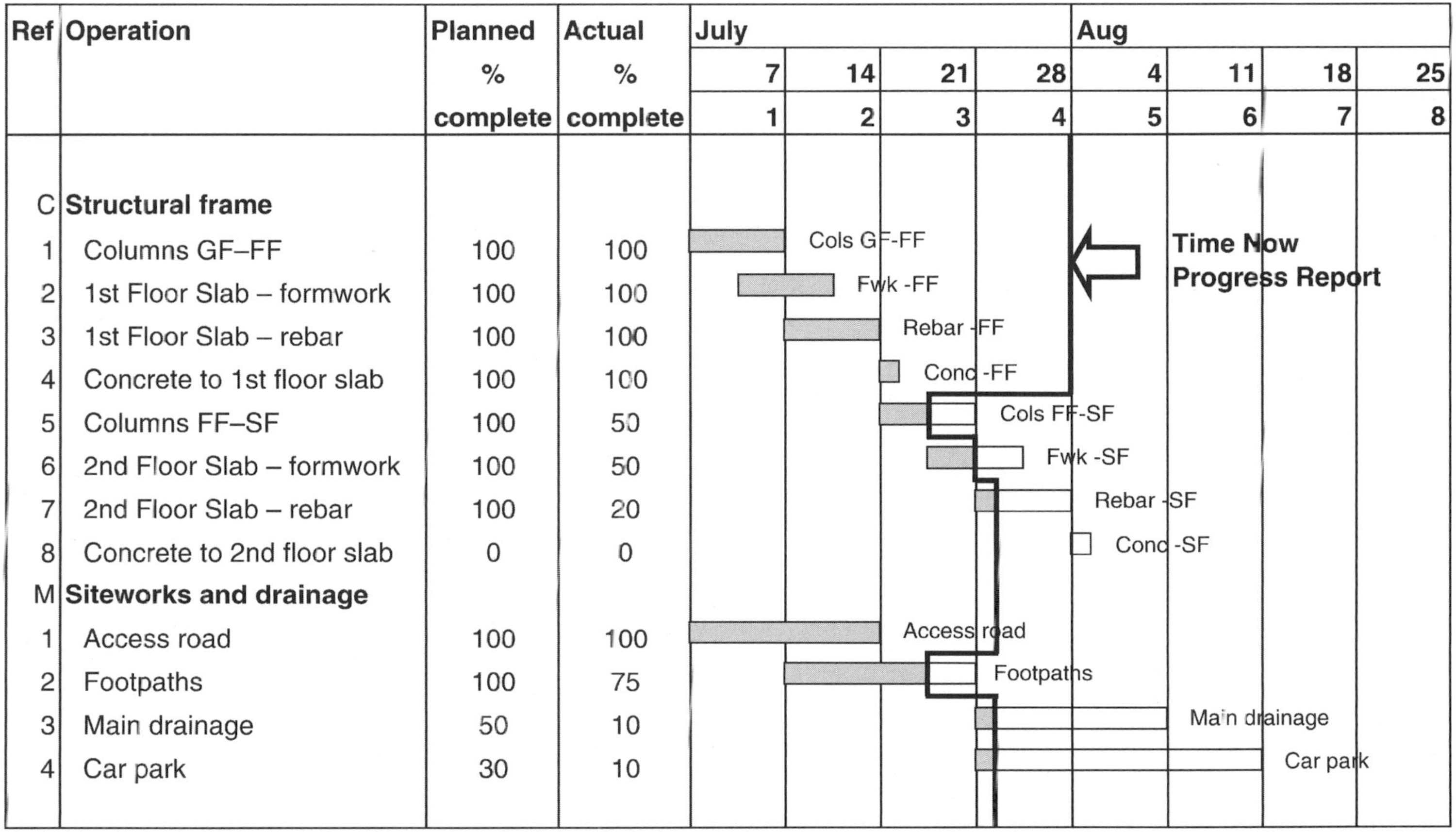

Ref	Operation	Planned % complete	Actual % complete
C	**Structural frame**		
1	Columns GF–FF	100	100
2	1st Floor Slab – formwork	100	100
3	1st Floor Slab – rebar	100	100
4	Concrete to 1st floor slab	100	100
5	Columns FF–SF	100	50
6	2nd Floor Slab – formwork	100	50
7	2nd Floor Slab – rebar	100	20
8	Concrete to 2nd floor slab	0	0
M	**Siteworks and drainage**		
1	Access road	100	100
2	Footpaths	100	75
3	Main drainage	50	10
4	Car park	30	10

Time scale: July 7 (1), 14 (2), 21 (3), 28 (4); Aug 4 (5), 11 (6), 18 (7), 25 (8)

Chart annotations: Time Now / Progress Report; Cols GF-FF; Fwk -FF; Rebar -FF; Conc -FF; Cols FF-SF; Fwk -SF; Rebar -SF; Conc -SF; Access road; Footpaths; Main drainage; Car park

Figure 9.2

PROGRESS CONTROL USING VALUE ENVELOPE

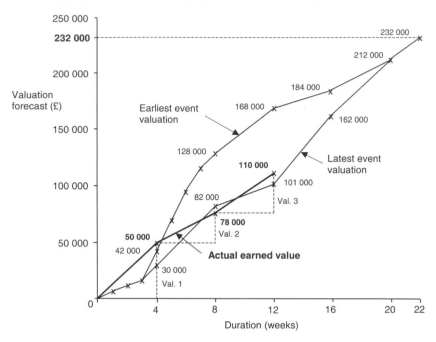

Figure 9.3

	VALUE FORECAST		ACTUAL EARNED VALUE	VALUATION PROGRESS REPORT
Week	Earliest Event	Latest Event		
4	42 000	30 000	50 000	Ahead of programme
8	128 000	82 000	78 000	Behind programme
12	168 000	101 000	110 000	On programme
16	184 000	162 000		
20	212 000	212 000		
22	232 000	232 000		

Figure 9.3 shows an example where the earned value is above the envelope in month 1 (generating value faster, therefore ahead), below the envelope in month 2 (behind) and inside the envelope in month 3 (on programme).

The monthly site progress report

It is common practice for an operational progress report to accompany the marked up bar chart. The progress report is normally written by the site planning engineer or the site manager. The report highlights the progress position operation-by-operation. Reasons for project delays may be shown. The current position in relation to key subcontractors, materials and orders yet to be placed is often highlighted.

An extract from a monthly site report in relation to Case Study No. 1 is given in Chapter 10.

9.5 Delay and disruption

Delay and disruption are common features of construction projects and this was highlighted in both the Latham and Egan reports.

Delay occurs when programmed activities do not start or finish on time, or both. If the contractor is delayed through no fault of his own, this will not necessarily result in an extension of time unless the contract completion date is affected. Delays to non-critical activities might result in a claim for loss and expense but there may be no entitlement to an extension of time because the completion date is not affected.

Disruption is where the contractor is involved in duplication of effort, out of sequence working or where he has to bring subcontractors back to site to do more work. Disruption to site activities may not lead to a delay but invariably will result in a claim for loss and expense.

The reasons for delays and disruption include:

- Incomplete design at commencement on site
- Variations and changes in the scope of the works
- Discrepancies in contract documents
- Unforeseen physical conditions
- Poor quality of workmanship
- Inadequate planning
- Underresourcing of site operations
- Accidents and incidents on site
- Bad weather

There are two types of delay:

- Culpable delay where the delay is the fault of the contractor
- Non-culpable delay where the contractor is delayed due to the fault of the client (or his agents)

9.6 Extensions of time

Standard forms of contract allow for extensions of time to be granted for delay which is not the fault of the contractor. These contractual provisions vary from contract to contract, as shown in the following examples.

JCT 98 Standard Form

Extensions of time are granted when the architect or contract administrator decides that a 'relevant event' has occurred which will delay the contract completion date.

Clause 25 of the contract lists 18 relevant events, including:

- Exceptionally adverse weather conditions
- Compliance with architect's instructions
- Failure of the architect to comply with an information release schedule
- Failure by the employer to give access to or egress from the site in time

A separate clause (clause 34, Antiquities) allows for giving extensions of time where special objects of interest are found.

JCT 98 With Contractor's Design

The provisions of this contract are similar to those of the Standard Form. There are 18 relevant events listed but some of these are slightly different. For example, clause 25.4.13 allows the contractor an extension of time when a change in the law alters the contractor's proposals and hence his design.

ICE Conditions 7th Edition

Clause 44 lists six grounds on which the contractor can apply for an extension of time:

- Variations ordered by the engineer
- Increased quantities
- Any cause of delay specified in the conditions of contract
- Exceptional adverse weather conditions
- Delay or default of the employer
- Other special circumstances (unspecified)

Possible causes of delay to be found in other clauses include:

- Clause 7 – delay in the issue of drawings and other instructions
- Clause 12 – adverse physical conditions or artificial obstructions
- Clause 40 – suspension of work
- Clause 42 – failure to give possession

Engineering and Construction Contract

Time is dealt with under core clause 3 but this mainly concerns the use and updating of the programme as a management tool, possession of the site, instructions to start or stop work and acceleration.

Extension of time provisions are covered under core clause 6 – Compensation Events. The clause defines 18 compensation events which include:

- Project manager's instructions
- Employer's failure to give possession
- Project manager's failure to reply to communications in time
- The contractor encounters physical conditions
- An employer risk event occurs

Where there is a single cause of delay there is usually little problem in dealing with prolongation claims from the contractor. However, in practice the situation is usually far more complex as several causes of delay frequently occur at the same time. Where there is concurrent delay (i.e. where the contractor is also in delay at the same time) the contractor's entitlement will be affected. Therefore, extensions of time for instructions issued by the architect after the contract completion date when the contractor is in culpable delay must be calculated on a net basis following the judgment in *Balfour Beatty Building Ltd* v. *Chestermount Properties Ltd* (1993).

9.7 Project acceleration

Where a project has been delayed for one reason or another, the contractor may wish to bring forward the completion date or may be asked to do so by the client (employer). If the contractor is not at fault in these circumstances, he should be paid for his trouble and effort through a separate agreement with the client. Some standard conditions of contract, such as the ICE Conditions, provide for this.

There are several types of acceleration, which have different legal meanings:

- **Pure acceleration** – where the contractor speeds up work on site so as to finish earlier than scheduled at the request of the client
- **Constructive acceleration** – where the contractor is effectively forced to work at a faster rate because the contract administrator has delayed or refused a legitimate application for an extension of time
- **Expedite** – where there is culpable delay and the pace of work on site has to be speeded up so as to get back on programme (see ICE Conditions for example)

Acceleration, or speeding up the work, can be achieved in three main ways:

- Re-organising the work more efficiently
- Increasing the resources on site
- Both

The work may be reorganised, for instance, by increasing the concurrency of site operations. This may be successful or may introduce other problems – inefficiencies or accidents for instance – where operatives and package contractors may be working on top of each other.

By increasing resources, work may be speeded up initially but here again problems may arise. For example, inefficient gang sizings may be introduced or the site may become congested with plant and operatives. There may also be problems of quality and supervision and extra costs may be incurred due to the need for additional site management personnel.

The contractor, therefore, needs to balance these and other considerations and decide to what extent it is physically and financially viable to speed up the work. There is a break-even point here beyond which diminishing returns may set in.

One method of seeking out this break-even point is to use the management technique of time–cost optimisation. This attempts to balance the direct cost of doing the work with the indirect costs of managing the process. The resultant analysis provides an optimum time and cost solution. The principles of time–cost optimisation, together with worked examples, are covered in Chapter 4.

9.8 Early warning systems

The success of a project is often related to the links developed between the main suppliers, subcontractors and the contractor. Bar chart displays which represent the connection between procurement and the commencement of operations on site have now become an integral part of planning. This is in order to ensure that key dates are met with respect to design requirements, information flow and the delivery or commencement of works on site.

Early warning systems were developed in the mid 1960s by John Laing, and extensively used. The system worked because of its simplicity in application and ease of monitoring; it allowed the user to develop his own symbols and letters to denote information requirements.

The NEC Engineering and Construction Contract (clause 1 – General) highlights requirements for early warning. This places responsibility on both the contractor and project manager to give an early warning

EARLY WARNING SYSTEM

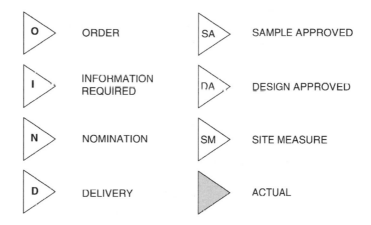

O ▷	ORDER
I ▷	INFORMATION REQUIRED
N ▷	NOMINATION
D ▷	DELIVERY

SA ▷	SAMPLE APPROVED
DA ▷	DESIGN APPROVED
SM ▷	SITE MEASURE
▶	ACTUAL

EARLY WARNING SYMBOLS FOR USE ON BAR CHARTS

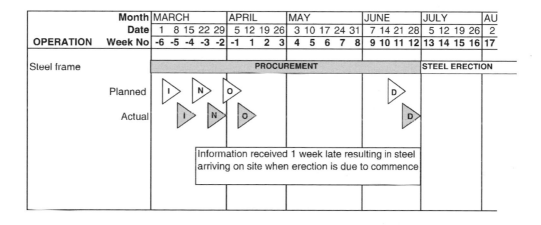

OPERATION	Month	MARCH					APRIL				MAY					JUNE				JULY				AU
	Date	1	8	15	22	29	5	12	19	26	3	10	17	24	31	7	14	21	28	5	12	19	26	2
	Week No	-6	-5	-4	-3	-2	-1	1	2	3	4	5	6	7	8	9	10	11	12	13	14	15	16	17

Steel frame — PROCUREMENT — STEEL ERECTION

Planned: I▷ N▷ O▷ ... D▷

Actual: I▶ N▶ O▶ ... D▶

Information received 1 week late resulting in steel arriving on site when erection is due to commence

EXAMPLE BAR CHART DISPLAY

Figure 9.4

on any matter which could delay completion or impair performance of the works in use.

Figure 9.4 shows a range of early warning symbols (or milestone symbols) which may be used to denote an occurrence which affects a supply chain for a component supplier or subcontractor. The example indicates that actual data was released by the client's representative 1 week later than planned in respect of information and nomination. This resulted in the order being placed 1 week later than scheduled with a possible delay in the commencement of steel erection on site. Information requirements relative to a project may also be presented in a tabular format, which meets the requirements of the JCT contract under clause 5.4.2. Delays in the release of information by the architect or other consultants may result in the contractor applying for an extension of time.

9.9 The control of money

The control of money in contracting and development is a complex subject but may be simply considered under the headings shown in Table 9.1.

Table 9.1 Control of money.

Cost	Cost = **money out** Wages, salaries, general costs and moneys paid or owing to builders' merchants, suppliers and trade subcontractors	Control must be exercised over **creditors** (to whom money is **owed**) by checking invoices carefully, making sure that the goods or services have been supplied and not paying the account too early
Value	Value = **money in** The value of work carried out as certified usually based on the quantity surveyor's interim valuation	Control must be exercised over **debtors** (from whom money is **owing**) to maximise **cash flow**. 'Value' excludes the value of **work in progress** (work carried out which has not been invoiced)
Cash flow	The **difference** between **money in** and **money out**	Sufficient **working capital** must be available to avoid **overtrading** (taking on too much work with insufficient liquid funds) and liquidity problems

9.10 Reporting procedures

Reporting procedures on the financial position of a contract must facilitate comparison of what is actually happening during the project with what was planned. It is essential that realistic budgets are prepared at the pre-contract stage and that they are monitored during the progress of the contract.

Monitoring the forecast during the project is an essential part of the project control process. The procedure for preparing cumulative value forecasts based on the master programme is outlined in Chapter 6 and illustrated in Figure 6.9.

Small firms

Within the small contracting organisation little is done regarding cost and value reporting during the project. The principal is more concerned about his cash flow position and keeping his bank at bay. Cash flow is vital for the survival of the small firm because of the shortage of working capital and the pressures coming from the bank and creditors (including the Inland Revenue and Customs and Excise).

The principal must keep the money flowing despite whatever crisis he may have on the project. Comparison of value with cost is only really made at defined stages of the contract, if at all. It may be undertaken at the completion of the substructure, superstructure work, or at completion of the building.

The majority of small contractors do not operate any form of cost control system. They may simply look at the actual cost and value situation, perhaps at the final account settlement stage, or they may be content to have their overdraft under control. This rather ad hoc approach to the management of cost information may be satisfactory for the small contractor, but as the business expands there is a greater need to report in a more formal manner.

Medium-sized firms

Within the medium-sized organisation, which may be undertaking contracts in the order of £500 000 in value, the reporting of financial information takes on more importance. Projects of this size cannot be allowed to drift along with management unaware of the financial position, and procedures must be implemented for reporting on contract profitability as the work proceeds.

The majority of medium-sized contractors undertake some degree of cost reporting during the progress of work. They also have cost–value reconciliation procedures as an integral part of the monthly valuation process, and these are conducted as one of the surveyor's normal monthly responsibilities. Whether or not senior management seriously consider the information to be reliable is another matter. A detailed account of the principles and practice of cost–value reconciliation procedures is described by Barrett (1981).

Large firms

As a company expands further, emphasis on the analysis of project performance becomes more important and consideration has to be given to more reliable methods of reporting. Within large organisations there is also a legal requirement for formal reporting procedures.

In the 1970s, a number of large construction firms operated reporting systems which collected data on every single site operation in progress. This practice appears to be out of favour these days and a more global approach is taken to the collection and analysis of data. Companies have at last begun to realise that collecting data for data's sake is an expensive luxury they simply cannot afford.

Financial control systems

As a general rule, 80% of the value of a contract is contained within 20% of the items in the bills of quantities. This is known as the Pareto principle. It is therefore worth considering setting up a control system to deal with the 20% of bill items which ultimately affect the contract's profitability.

Changes in the industry resulting from the introduction of information technology, the greater use of subcontractors and the use of work packages have resulted in the collection and analysis of more selective data. Reductions in profit margins and staffing levels have led to a serious review of existing control systems in order to make the business a little leaner and to gain a more competitive edge. Old, outdated systems have been abandoned due to changes in the nature of construction work practices.

9.11 Monthly cost–value reporting

Cost–value reconciliation (CVR) is the comparison of the project value with the project cost at predetermined intervals during the progress of a project. This interval is normally monthly and tends to tie in with the company's valuation and accounting procedures. The purpose of CVR is to allow management and statutory accounts to be prepared on a more meaningful basis. The basic principles of cost–value reconciliation are illustrated in Figure 9.5.

It is worth noting that a new approach to understanding accounts has been suggested by a number of Manchester academics as part of a new accounting system which attempts to put a 'brave face' on company figures. It has been devised as a way of assessing a company's health by using cartoon faces. This approach should appeal to those responsible for interpreting cost–value reports in construction and should assist in providing instant feedback to senior management. Figure 9.6 illustrates the possible application of 'expression management' to cost–value reporting.

PRINCIPLES OF COST–VALUE REPORTING

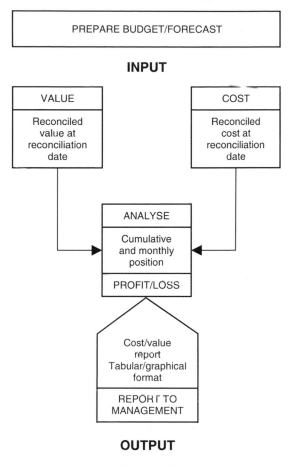

Figure 9.5

9.12 Cost–value reconciliation terminology

The procedure for reconciling cost and value at a cut-off date at the end of each month during a project is outlined in Figure 9.7 and this should be read in conjunction with the following definitions.

Forecast value (cumulative)

This is a forecast of the cumulative value based on the contract master programme. It is obtained by allocating money to the bar chart or schedule of operations and presenting the monetary figures in the form of a cumulative line graph or cumulative value forecast. It must, however,

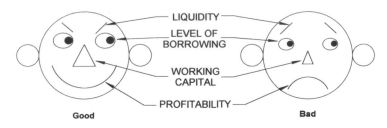

Good

Bad

EXPRESSION MANAGEMENT

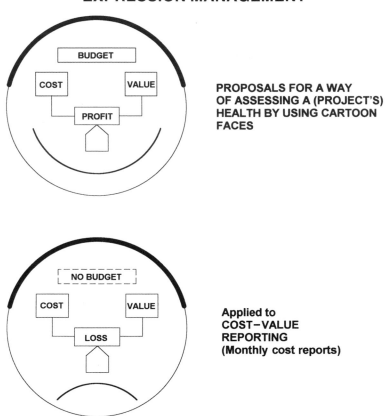

PROPOSALS FOR A WAY
OF ASSESSING A (PROJECT'S)
HEALTH BY USING CARTOON
FACES

Applied to
COST–VALUE
REPORTING
(Monthly cost reports)

Figure 9.6

be based on a realistic assessment of the sequence of work or contract programme (refer to Chapter 6, Figure 6.6). The forecast may be presented in tabular or graphical format.

Reconciled value (cumulative)

This is the project value assessed at the cut-off date, which is often referred to as the contract 'cost–value reconciliation date'. This may be

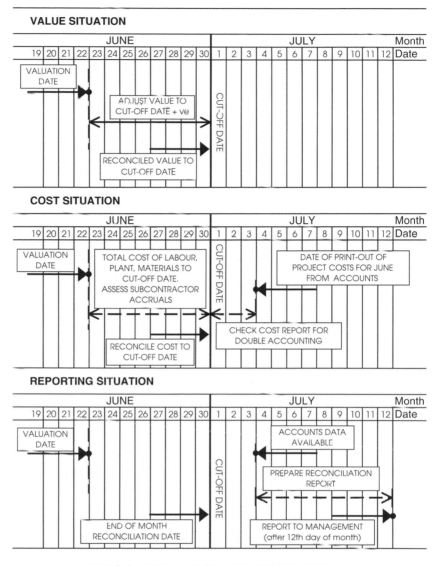

RECONCILIATION PROCEDURE

Figure 9.7

somewhat different to the value agreed at the monthly valuation date with the professional quantity surveyor and the contractor. This value, for the purposes of comparison with the project costs, must be the reconciled value adjusted for the time variance between the valuation date and the reconciliation date. This may involve assessing the value of the work undertaken in the intervening period between the actual valuation date and the cut-off date. Depending on valuation date, the

ASSESSMENT OF RECONCILED VALUE

GROSS VALUE		VAL. DATE 23 May		
Gross value certified by the professional quantity surveyor				176 000
		ADD	OMIT	
ADJUSTMENTS TO RECONCILIATION DATE 31 May				
PLUS or MINUS				
Undervaluation +ve				
Overvaluation −ve			5 000	
Adjusted value to date of reconciliation +ve		15 000		
Variations issued (not yet included in valuation)		2 000		
Dayworks – ditto.		1 000		
Remeasured work sections		3 000		
Preliminaries adjustments – undervaluation +ve or overvaluation −ve			2 000	
Materials on site adjustments		1 000		
		22 000	7 000	15 000
RECONCILED VALUE				**£191 000**

Figure 9.8

adjustment may be positive or negative. Figure 9.8 indicates a typical format for presenting the assessment of the reconciled value.

Reconciled cost (cumulative)

This is the cost expended at the date of reconciliation. It is the cost in the contractor's cost ledger, adjusted to the cut-off date or reconciliation date. The cost assessment must include all accruals for materials and subcontractors which have not been included in the cost ledger at the date it was closed. Assessment of the reconciled cost is one of the major areas of error in the reconciliation process, especially in the assessment of the subcontractor accruals. Similar errors also occur in the assessment of the material costs. Materials accruals represent the cost of materials delivered to site but not yet included in the cost ledger. For this purpose it may be necessary to put a value to goods received records and material delivery notes.

Reconciliation date (or cost–value cut-off date)

This is the date agreed by management when the comparison or reconciliation of cost and value is to take place. It is usually the date when the monthly accounting period is closed and is frequently referred to as the cut-off date. This may be the last Friday in each month, the 30th of each month or simply the last day in the month.

Date of report to management

The report date in principle should be as close to the cut-off date as possible but will depend on when the project cost figures are available. Figure 9.7 illustrates this as being achieved by the end of the 12th day of the month.

9.13 Cost–value reports

Figure 9.9 illustrates a typical format for a cost–value reconciliation report in a large contracting organisation. Various adjustments to the cost are indicated in order to match cost with the adjusted value.

Accruals are costs which have been incurred but for which invoices have not been received by the accounts department. Accruals represent provisions recognising future costs which have to be assessed and taken into account in order to report true value.

The comparison of the cumulative value with the cumulative cost is often referred to as the monthly cost report and a suitable form for this is shown in Figure 9.10. The comparison may be presented in such a table or alternatively in graphical format. Figure 9.11 indicates the principles of the relationship between the cumulative value forecast, actual reconciled value and reconciled cost presented in graph form. The value variance and time variance have been highlighted.

Variance analysis which highlights the difference between actual and expected figures forms an essential part of the cost–value reporting procedure, but it should be noted that it is important to review the contract's progress when considering variances as there must be reasons for any shortfall in the project value.

Data relating to a project has been indicated on the monthly cost report in Figure 9.10, together with the forecast cumulative value for the project. The relationship between forecast value, actual value and cost has been presented in graphical form in Figure 9.12. The percentage

COST–VALUE RECONCILIATION REPORT		
Contract _____	Valuation No. _____	
Contract No. _____	Date of valuation _____	
Contract duration _____	Month _____	
VALUATION ASSESSMENT	**CUMULATIVE**	**THIS MONTH**
Value of certificate to / /		
ADJUSTMENTS :		
Adjustment to valuation date		
Preliminaries adjustment		
Overvaluation		
Variations		
ADJUSTED VALUATION TOTAL		
CONTRACT COST ASSESSMENT	**CUMULATIVE**	**THIS MONTH**
Contract costs to / /		
ADJUSTMENTS TO COST (ACCRUALS)		
Plant		
Materials		
Subcontractors		
Inter-site costs		
PROVISIONS		
Subcontractor liabilities		
Future losses		
Maintenance/defects costs		
Cost of delays		
Liquidated damages		
ADJUSTED COST TOTAL		
PROFIT (LOSS) As a value		
Percentage		
Date of reconciliation / /	Prepared by :	

Figure 9.9

profit release situation has been displayed in both cumulative and monthly terms in Figure 9.13.

Management tends to react more to data presented graphically, which clearly indicates the relationship between forecast, actual value and cost.

MONTHLY COST REPORT
COST–VALUE RECONCILIATION

Date	Val. No.	Certified value	Recon. value	Cum. cost	Cumulative Profit	%	Monthly Value	Cost	Profit	%
30/5	1	27 000	29 000	25 000	4 000	16.0	29 000	25 000	4 000	16.0
30/6	2	64 000	67 000	59 000	8 000	13.5	38 000	34 000	4 000	11.7
28/7	3	110 000	112 000	102 000	10 000	9.8	45 000	43 000	2 000	4.6
26/8	4	170 000	175 000	163 000	12 000	7.4	63 000	61 000	2 000	3.3
25/9	5	270 000	280 000	265 000	15 000	5.6	105 000	102 000	3 000	2.9

Forecast cumulative value
(based on programme)

Val. No.	Forecast Cumulative Value	Val. No.	Forecast Cumulative Value
1	30 000	5	310 000
2	70 000	6	410 000
3	130 000	7	480 000
4	200 000	8	560 000

Figure 9.10

Commentary on Figure 9.13

The cumulative profit release has been slowly declining each month of the project. In 5 months profit has fallen from 16% to 5.6%. During this period, the monthly profit has declined from 16% to 2.9%. During months 3, 4 and 5 the average monthly profit release has been some 3.5%. At month 5 of the 8-month contract, the cumulative profit release of 5.6% is well below the forecast profit of 10%. It is doubtful that the

PRINCIPLES OF COST–VALUE RECONCILIATION ANALYSIS PRESENTED GRAPHICALLY

Value–time/Cost–time relationship

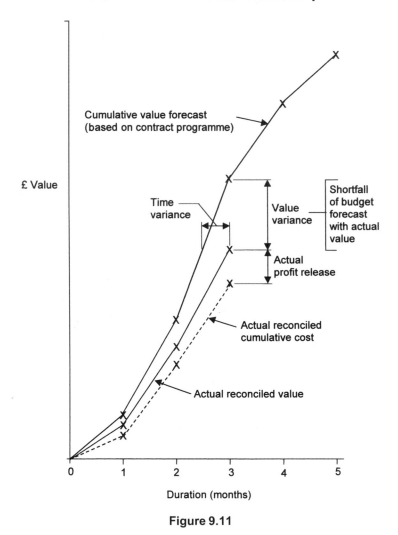

Figure 9.11

contract will achieve its forecast margin without some drastic action by senior management.

Part of the cost–value reporting procedures within the larger contracting organisations includes preparing a forecast of the project's cost and profitability to completion. This will involve consideration of the profit level capable of being achieved on the remaining operations to be completed (between month 5 and the end of the contract).

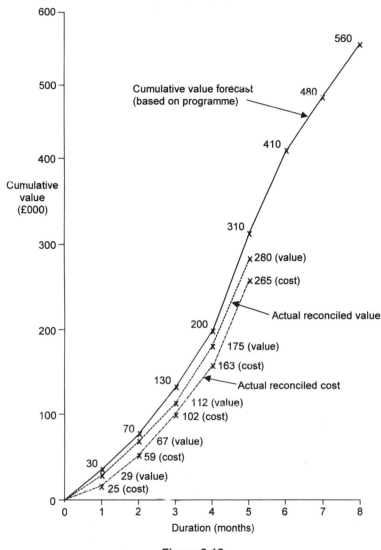

VALUE–TIME/COST–TIME RELATIONSHIP

Figure 9.12

9.14 Cost–value reconciliation case study

Contract brief

Table 9.2 indicates the contract value and cost position at the end of month 3 of a 4-month project.

The forecast project profit margin is 10%.

PERCENTAGE PROFIT RELEASE

Date	Validation No.	CUMULATIVE		MONTHLY	
		PROFIT	%	PROFIT	%
30/5	1	4 000	16	4 000	16
30/6	2	8 000	13.5	4 000	11.7
28/7	3	10 000	9.8	2 000	4.6
26/8	4	12 000	7.4	2 000	3.3
25/9	5	15 000	5.6	3 000	2.9

Profit forecast at tender stage 10%

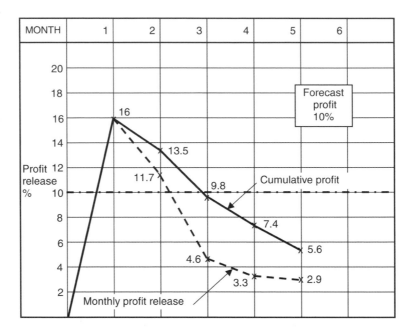

Figure 9.13

Table 9.2 Contract value and cost position.

Month	Cumulative value forecast (£)	Actual cumulative value (£)	Actual cumulative cost (£)
1	30 000	22 000	20 000
2	90 000	65 000	60 000
3	140 000	100 000	92 000
4	180 000		

Analysis of the variance at the end of each month

An analysis of the variance at the end of months 1 to 3 has been outlined in the form of a report to senior management, as follows.

Contract variance analysis at the end of month 1			
	£		£
Forecast value	30 000	Actual value	22 000
Actual value	22 000	Actual cost	20 000
Value variance	**8 000**	Profit variance	**2 000**

The value variance of £8000 represents a time variance of approximately 1 week delay in the progress of the works. This position must be verified by contract progress reports and the reasons for the 1-week delay at this stage of the project must be established. Is the delay due to the contractor, his subcontractors or delay in the receipt of information from the architect? All these questions must be answered.

The actual profit release to date is 10% and the contract is therefore within the forecast margin.

Contract variance analysis at the end of month 2			
	£		£
Forecast value	90 000	Actual value	65 000
Actual value	65 000	Actual cost	60 000
Value variance	**25 000**	Profit variance	**5 000**

The value variance of £25 000 represents a time variance of approximately 2 weeks delay in the progress of the works. The reasons for the 2-week delay must again be investigated and actual progress checked against the master programme. The effect on the project completion date must also be assessed and, if appropriate, requests for an extension of time should be considered.

The cumulative profit release to the end of month 2 is 8.3% and the monthly profit release is 7.5%. Both of these figures indicate a need for concern at this stage of the project.

It is recommended that the work rate be increased during the next 8 weeks in order to bring the contract back on programme. Intensive short-term planning procedures should be implemented at site level.

Contract variance analysis at the end of month 3			
	£		£
Forecast value	140 000	Actual value	100 000
Actual value	100 000	Actual cost	92 000
Value variance	**40 000**	Profit variance	**8 000**

The value variance of £40 000 is now giving rise for concern as the project is now some 3 weeks behind programme. In order to complete the programme on time some £80 000 value of work will require to be completed next month. Realistically this is not possible considering the project's past performance.

The action recommended at the end of month 2 has not been fully implemented. A serious overview of the reason for the delays to date must be undertaken. A realistic project completion date must now be assessed and the client informed of any delay to the project completion date.

The cumulative profit achieved to date is 8.7% and the monthly profit release is 9.4%. Possible liquidated damages to be levied at contract completion will affect the final profit release. A cost and value forecast to the completion of the project should be prepared.

Figure 9.14 indicates the relationship between forecast, actual value and actual cost presented graphically. At the end of month 3 the total value variance is £40 000 (negative) and the time variance shows the project approximately 3 weeks behind programme.

9.15 Interim valuations using S curves

Case study

Project particulars

- The construction of a 60-bed hotel and adjacent leisure facilities. The six-storey hotel building is of timber-framed construction with traditional brick finishes
- The contract has been awarded as a design and build project as part of a partnering arrangement with the contractor. To date 20 similar projects have been successfully completed under this arrangement
- Project value is some £2.1 million with a contract period of 28 weeks. The contract payment terms are every 2 weeks and payment is to be made within 21 days. The contract is fast track using named subcontractors, all of whom have worked with the contractor on previous similar contracts
- The project is under the control of an executive project manager employed by the client, and extensive trust has been developed between the parties over the partnering period

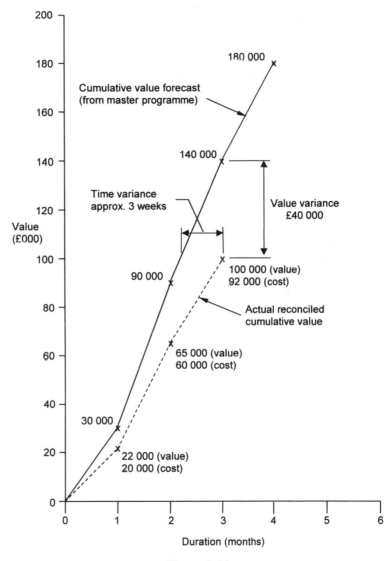

VARIANCE ANALYSIS

MONTHLY AND CUMULATIVE REPORT

Figure 9.14

The valuation process

Interim payments (every 2 weeks) are based on a forecast S curve agreed between the client and contractor at project commencement. The S curve is based on similar principles to formulae used by some government departments. An abstract from the S curve print-out is given in Table 9.3.

Table 9.3 Abstract from S curve.

Week No.	Cumulative value forecast (£)	Week No.	Cumulative value forecast (£)
2	69 250	16	1 311 000
4	180 600	18	1 506 000
6	326 500	20	1 683 000
8	499 300	22	1 835 000
10	691 800	24	1 953 000
12	896 200	26	2 031 000
14	1 105 115	28	2 100 000

Release of valuation moneys

The client's project manager allows the contractor to self-certify up to the end of week 14 (7th payment stage). At this point in the contract a check is made on the progress of the works. If the contract is on programme, self-certification continues up to week 24 (12th payment stage) of the project. At this stage a further progress check is made and the value of contract variations to date assessed and added to the next payment.

At the end of the contract, the contractor submits his final account within 6 weeks of contract handover.

Note: This is a practical example of partnering at its best – where complete trust exists between client, contractor and subcontractors.

9.16 The control of resources

The control of project resources requires consideration of:

- Labour
  - Directly employed labour
  - Subcontract labour
- Materials
  - Control of waste at site level
- Plant
- Preliminaries

Labour and materials control in particular is discussed below, while procedures for forecasting and monitoring of labour and plant are dealt with in Chapter 6. The control of subcontractors is discussed separately in Chapter 7.

9.17 Labour control

The widespread use of subcontractors in construction has shifted the emphasis away from the need for contractors to employ sophisticated labour control procedures, as was the case in the days of the general contractor who directly employed his own workforce.

On large projects, the contractor is more concerned with keeping the overall labour expenditure within the estimated allowance. The moni toring of labour expenditure involves recording the actual man-weeks expended each week on the project and matching this with the forecast. It is important to relate the analysis to the actual contract progress situation, as the reason for an apparent overspend can simply be due to the project being ahead of programme. See also Chapter 6.

9.18 Materials control

Materials expenditure represents a major proportion of contract value. Therefore, the control of purchasing, scheduling, delivery and handling of materials on site is an essential part of the control process.

Within small organisations, the responsibility for all aspects of material control lies with the principal. He is the estimator, buyer, surveyor, and contracts manager all rolled into one.

In medium-sized companies, responsibility for the purchase of materials may rest with the estimator/buyer or may be part of the surveying function. However, once materials have been delivered to site, they become the sole responsibility of the construction site manager.

Large companies, on the other hand, usually have departments responsible for buying and procurement, estimating, surveying, contracts and administration, all operating within their own little empires. Within such organisational structures, these separate functions operate more formally and therefore a formalised approach to the procurement and management of materials is necessary. In the worst possible scenario, communication problems between departments may arise where, for instance, the chief buyer is not talking to the chief estimator for some reason. Perhaps the contracts director thinks the chief surveyor is an idiot – and on it goes!

Good communications are important and the buying and contracts sections need to liaise closely in order to ensure that materials arrive on site on time. Contact needs to be established between the contract buyer and the site manager in order to ensure that the material call-off schedules clearly tie in with the programme of work.

Responsibilities also have to be clearly defined. The responsibility for buying materials within the estimate allowances lies with the buyer and any resulting savings created by efficient buying contribute to the profitability of the contract. The responsibility for handling materials, distribution around the site and fixing them in position belongs to the site manager, who is also responsible for material loss and accounting for excessive waste.

Procedures for reducing loss and waste on site will be reviewed later in this chapter.

A materials management system for speculative housing

Principles of the system

For a materials management system, the objective is to produce material schedules which will be of benefit to the site manager. The system should combine the material scheduling with the planning and control of the project. Material schedules have to be synchronised to the sequence of work (or elements of construction) on each phase of a project and also be directly related to the requirements of the contract programme. The site manager is responsible for calling-off or requesting delivery of materials to the site and for keeping suppliers informed of revised delivery dates in situations where work on site is behind or ahead of schedule.

Stage 1 – Establishing the construction elements

The stages of construction should be divided into elements, such as:

(1) Foundations
(2) Ground floor slab
(3) Internal and external walls
(4) Intermediate floors
(5) Roof construction
(6) First fix
(7) Second fix
(8) Final fix
(9) Sanitary appliances
(10) Drainage
(11) Paving and landscape
(12) Roadworks and main sewer work

Certain of these elements may not be applicable in cases where the work is to be sublet to a subcontractor who supplies materials. The elemental list may be extended to suit the requirements of individual projects.

Stage 2 – Establishing the materials schedule

For each of the elements indicated it is necessary to identify a schedule of materials. This will only relate to the materials to be scheduled and

MATERIAL SUMMARY

ELEMENT – 1	FOUNDATIONS		Contract: Date:	
REF.	MATERIAL	SUPPLIER	ORDER NUMBER	SCHEDULE REFERENCE
1/1	Ready mixed concrete			C1
1/2	Hardcore – brick			C2
1/3	Hardcore – stone			C3
1/4	Visqueen DPM			C4
1/5	Reinforcement fabric			M1
1/6	Reinforcement bar			M2
1/7	Common bricks			B1
1/8	Sand lime bricks			B2
1/9	Engineering bricks			B3
1/10	Facing bricks			B4
1/11	Trench blocks			B5
1/12	Blockwork			B6
1/13	Wall ties			B7
1/14	Air vents			B8
1/15				
1/16				
1/17				

Figure 9.15

ordered by the buying department. Figures 9.15 and 9.16 show a materials summary sheet for the foundations and external walls element.

The materials scheduler (working within the buying department) is responsible for taking-off quantities from the drawings and setting up the materials schedules for each element. Minor materials which are to be purchased at the local builders' merchants will not be included on the schedule. The key material delivery dates will be established from

MATERIAL SUMMARY

ELEMENT – 3	EXTERNAL WALLS		Contract: Date:	
REF.	MATERIAL	SUPPLIER	ORDER NUMBER	SCHEDULE REFERENCE
3/1	Facing bricks			B9
3/2	Facing bricks			B9A
3/3	Facing bricks			B9B
3/4	Common bricks			B10
3/5	Blockwork			B11
3/6	Blockwork			B12
3/7	Flue blocks			B13
3/8	Metal lintels			B14
3/9	Timber lintels			B15
3/10	PC. lintels			PC1
3/11	Ext. door frames			J1
3/12	Ext. windows			J2
3/13	Ext. windows			J2A
3/14	Special frames			J2B
3/15	Garage door frames			J2C
3/16	Steel beams			S1
3/17	PC. thresholds			PC2
3/18				
3/19				
3/20				

Note: The external wall element covers all items in the superstructure shell (i.e. external/internal walls)

Figure 9.16

the contract programme after consultation with the site manager and contracts manager. Figure 9.17 indicates the layout of the materials schedule.

The material control system outlined above is ideal when the contractor is building standard house types on different development projects. Where this is the case, the materials scheduler has only to prepare

ELEMENT 1 – FOUNDATIONS   MATERIALS SCHEDULE   SCHEDULE REF. B
MATERIAL – COMMON BRICKS   Ref. B1. B3. B5.

| Supplier | Address | Telephone number | Order number | Date order placed | Material description | Total quantity | CALL OFF DATES | | Notes |
							Date	Quantity	
Armstrong	84, Lord St.	0151-702	73496	12/2/97	Common	15,000	13/6/97	8,000	
Bricks	Southport	3849			Bricks		20/7/97	7,000	
	SK7 4RL								
Armstrong	Ditto	Ditto	73497	12/2/97	Engineering	5,000	13/6/97	5,000	
Bricks					Bricks				
Tarmac	97, Albert Rd.	0161-798	75001	15/3/97	Trench	15,000	2/6/97	5,000	
	Manchester	3026			Blocks		20/7/97	5,000	
	M21 3RU						15/8/97	5,000	

Note: Separate schedules may be provided for each material or
similar material may be combined on a single schedule

Figure 9.17

material requirements for each house type and then apply the schedules to the phasing of the project.

9.19 Materials waste on construction sites

A report published by the Institution of Civil Engineers (ICE 1996) indicates that over 500 million tonnes of construction waste are generated each year. Landfill space is becoming scarce and the cost of tipping materials is very high and rising. The introduction of the Landfill Tax in 1996 has trebled the cost of disposing of hired skips and the cost of this will ultimately be passed on to the client in the form of increased tender prices.

For all these reasons, therefore, waste management on construction sites should be taking on more importance.

Extensive research has been undertaken into the subject of material waste on construction sites by Skoyles and Hussey (1974) and several books and CIOB technical papers (CIOB, undated) have been published on the topic.

Many of these references are years old and so in order to establish some evidence of current practice in materials management for this book, we undertook a small survey. Eight housing construction sites were visited in the Manchester region, all of which were being developed by established housing contractors.

Although this evidence is anecdotal it is probably a fair reflection of the standards of materials management generally. Quite frankly, the standards were shocking. It appears from these observations that little has been learnt from the lessons of the past and that the site managers concerned were unaware of any materials policy within their organisation. Any regard at all for materials waste appeared to be entirely discretionary and on six of the eight projects materials were appallingly mismanaged. Illingworth (1993) makes similar observations based on personal experience of waste management.

Observations relating to the mismanagement of materials included:

- Excessive waste left under scaffolds, including bricks, blocks, skirting boards, fascia boards, drainage fittings, etc.
- Expensive facings and engineering bricks being bulldozed into the ground and then covered over with topsoil to provide 'instant brick gardens'
- Materials being stored on uneven ground, adjacent to unprepared access roads, allowing the material to become contaminated with mud and water
- Pallets of bricks and blocks unloaded directly on to unprepared ground, away from the workplace

- Damage to materials while unbanding the packs
- Roof trusses being stacked on unprepared areas, allowing them to distort and twist
- Lack of covering and protection to internal timber floor joists, door frames and finishing joinery items. Structural timbers left unprotected in the rain
- Excessive thickness of ready mixed concrete to in situ-concrete kerb beds
- Out of sequence working, resulting in the excessive waste of stone filling materials, bricks and blocks, etc.
- Commencing foundation work with no provision for adequate access to the works. This resulted in chaos with respect to the storage of materials around the work area

Only two of the companies had a materials management policy. On a number of projects, the site manager placed the blame for materials mismanagement on the extensive use of subcontract labour. Site managers also stated that they had far too much to do to worry about materials waste and were unaware of wastage allowances in the estimate.

Wastage allowances in the contractor's estimate

The contractor's estimator allows for material waste when building up the unit rates for materials at the estimate stage. Each estimator has a different perspective on waste allowances but Skoyles and Hussey (1974) found that the percentage allowances compared to actual waste for a variety of materials were as shown in Table 9.4.

Table 9.4 Allowances compared to actual waste.

Item	Location	Normal estimators allowance (%)	Typical loss in practice (%)
Ready-mixed concrete	In foundations	2.5	10
	In formwork	2.5	6–10
Bricks and blocks	Commons	4	8
	Facings	5	12
	Engineering	2.5	10
	Lightweight blocks	5	10
	Concrete blocks	5	7
Roofing	Tiles	2.5	10
	Felt	2.5	8
	Lead flashings	2.5	7
Joinery items	First fix timbers	5	10
	Boarding	5	15

Recommendations for reducing waste

Waste on construction sites is caused by a number of factors which may be conveniently categorised under four headings:

- Design waste – where the design results in wasteful cutting on site
- Take off/specification waste – where additional materials are ordered 'to be on the safe side' or where materials are over-specified for the job in hand
- Delivery waste – materials damaged in transit or during off-loading
- Site waste – waste resulting from the production process or where incorrect materials are used

Delivery and site waste can be reduced by considering the following:

- Allow adequate moneys in the estimate for the provision of temporary access roads and hardstanding areas for plant and equipment. Allow for providing material storage areas and facilities to store materials clear of the ground (storage racks)
- Ensure good site layout planning at the pre-contract stage. Allow the site manager to become involved in the decision-making at this stage of the project
- Encourage the site manager to manage his materials properly – he can be assisted by establishing in-company courses on materials management. The company may offer some degree of incentive to managers who achieve the minimum waste targets on their projects
- Ensure that the staff are aware of the company materials management policy and of the allowances built into the estimate
- Make senior managers responsible for the losses occurring on their projects as well. It is no use blaming the site manager when his superior is oblivious to the losses when he visits the site
- Ensure adequate control procedures for receiving materials on site including checking of delivery notes, correct handling and protection of materials on site. Ensure the use of correct lifting equipment which does not damage the goods. Provide adequate facilities for the unbanding of material packs
- Provide adequate site security facilities and materials storage compounds
- Provide simple procedures and checks for reconciling materials utilisation during the project. This could form part of the surveyor's role at the valuation stages
- Make subcontractors aware of the company's materials waste policy. Establish procedures for contra charging subcontractors for the excessive waste they create

The company may consider producing a simple site guide on its waste management policy. Recommendations may be included on good practice regarding materials handling and control. This could be issued to subcontractors and the company's directly employed labour.

Also, poor control of material waste can create health and safety problems. Procedures should be included in the contractor's safety management system to eliminate this potential hazard.

9.20 Key performance indicators

In order to measure overall performance on projects, key performance indicators (KPIs) may be used. They can be used to set standards at the start of a project, monitor performance as the project progresses and form a basis for improvement targets for future projects.

Through 'Rethinking Construction' and the Construction Best Practice Programme (merged into 'Constructing Excellence' in 2004 – www.constructingexcellence.org.uk) initiatives set up by the Government to assist the construction industry have resulted in the development of KPIs for several aspects of project delivery including:

- Client satisfaction
- Defects
- Safety
- Time and cost predictability
- Productivity
- Profitability

By using KPIs, companies can measure project performance both internally and externally and thereby establish benchmarks to help the organisation achieve best practice standards.

References

Barrett, F. R. (1981) *Cost Value Reconciliation*. Chartered Institute of Building.

CIOB. Technical Papers 1, 15, 60, 87 and 92 on loss and waste. Chartered Institute of Building.

ICE (1996) *Managing and Minimising Construction Waste*. Institution of Civil Engineers.

Illingworth, J. (1993) *Construction Methods and Planning*. E. & F. N. Spon.

Skoyles, E. R. & Hussey, H. J. (1974) Wastage of materials. *Building*, Feb.

Thomas, R. (2001) *Construction Contract Claims*, 2nd edn, Palgrave.

10 Planning case studies

10.1 Introduction

The objective of this chapter is to provide the reader with the opportunity to follow a series of worked examples relating to a variety of different construction projects in order to build confidence and understanding in the application of the various techniques referred to in this book.

This is done using five case studies which encompass both building and civil engineering work. The case studies and the worked examples are summarised below.

10.2 Summary of case studies

Case study No. 1 – City Road project

Project

A £14 million multi-storey development of luxury apartments.

Worked examples

(1) Tender stage
    (1.1) Pre-tender programme
    (1.2) Build-up of site management costs (preliminaries)
    (1.3) Method statement for basement excavation
    (1.4) Cranage strategy
(2) Pre-contract stage
    (2.1) Site layout planning and materials handling
    (2.2) Sequence study for basement construction
    (2.3) Project quality plan
(3) Contract stage
    (3.1) Subcontract procurement and management procedures
    (3.2) Procedures manual for planning and control
    (3.3) Progress records and reports

Case study No. 2 – Shopping mall project

Project

£300 million shopping mall complex.

Worked examples

(1) Project relationships
(2) Project planning and control strategy

Case study No. 3 – Housing project

Project

Construction of 2 eight-storey reinforced concrete-framed buildings and 28 low rise housing units.

Worked examples

(1) Construction strategy
(2) Tender method statement
(3) Pre-tender programme

Case study No. 4 – Canal project

Project

£1 million canal refurbishment project.

Worked examples

(1) Method statement
(2) Clause 14 programme
(3) Risk assessment
(4) Safety method statement

Case study No. 5 – Highway project

Project

Major highway project.

Worked examples

(1) Construction strategy

(2)   Method statement

(3)   Clause 14 programme

General note

The following case study examples reflect the key problem areas encountered by the contractor's team at both the tender and pre-contract stages of the project. Data has been collected from discussions with the tender team, chief planning engineer and project manager in order to record actual decisions made at each stage of the project.

10.3 Case study No. 1 – City Road project

Project description

This case study is based on the City Road project which is a development of three blocks of luxury riverside apartments complete with private basement car parking, roadworks and landscaping.

The project involves the construction of 3 seven-storey buildings on a restricted site in the centre of a major city in the north-west of England. A two-level basement (Basement B) is to be constructed under Block 1. Blocks 2 and 3 are to be constructed over a single-level basement (Basement A). The multi-storey blocks are of steel frame construction incorporating metal deck floors and brick external facades.

The site is bounded on the north side by the River Med where there is an existing retaining wall which is to be demolished and rebuilt. The south side of the site is bounded by City Road which is a busy main thoroughfare.

The general layout of the project is shown on the site location plan (Figure 10.1) and Figure 10.6 shows the basement location plan.

Project details

Contract value is £14 million and the agreed contract period is 70 weeks. The contract agreement is to be executed under the JCT 98 with Contractor's Design form and the employer's scope designers are to be novated to the contractor to create a single point of responsibility for design and construction.

The relationship between the client's design team and the contractor is shown in Figure 10.2.

Buildability is seen as an important issue on this project and the contractor has identified a number of key areas which warrant particular

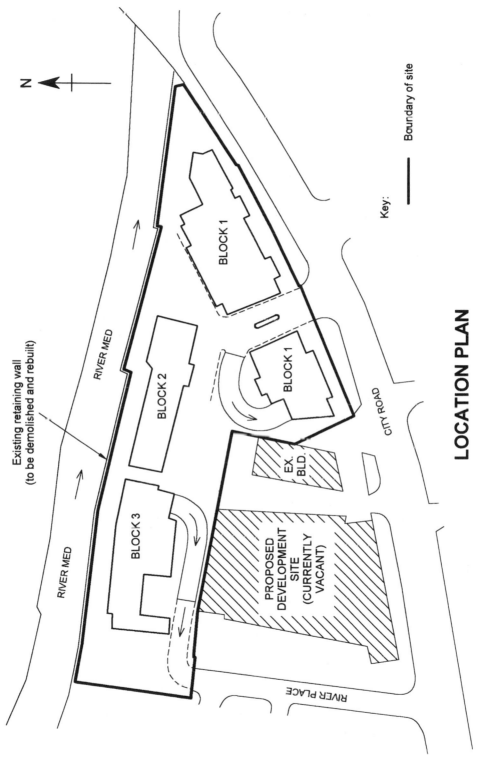

LOCATION PLAN

Figure 10.1

Key: —— Boundary of site

BLOCK 1

BLOCK 1

BLOCK 2

BLOCK 3

RIVER MED

RIVER MED

Existing retaining wall
(to be demolished and rebuilt)

CITY ROAD

RIVER PLACE

EX. BLD.

PROPOSED
DEVELOPMENT
SITE
(CURRENTLY
VACANT)

N

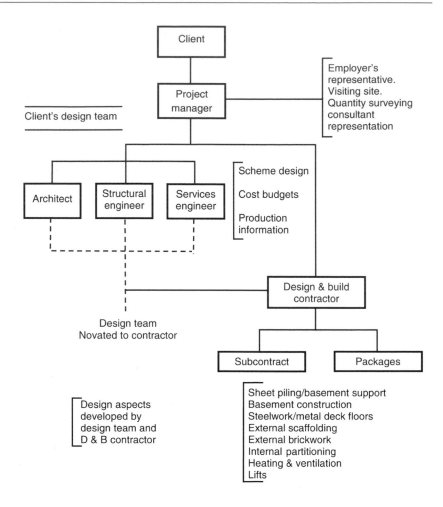

**CLIENT, DESIGN TEAM, CONTRACTOR
RELATIONSHIP, CITY ROAD PROJECT**

Figure 10.2

attention, including basement construction proposals and the steel frame erection sequence.

The contractor proposes to let out the work as a series of work packages and some of these will be delegated with design responsibility.

Site constraints

The following factors are considered important in developing the construction methodology and sequence of work:

- The only access in and out of the site is via River Place
- No access is available on the north side of the site due to the presence of the River Med
- Limited access is available on the south side due to City Road.
- Working space on site is limited and there is no space available adjacent to the site due to adjoining developments
- Limited site space for the storage of materials
- Basement B under Block 1 is 6 m deep and is located at the rear of the footpath on City Road
- Extensive work is to be undertaken in rebuilding the river retaining wall. This work is to be completed in sequence with the basement construction under Blocks 2 and 3 (Basement A)
- There are legal problems concerning cranes oversailing adjacent properties and the client might be asked to negotiate an easement with adjoining owners. Alternatively, luffing jib cranes could be used which can raise their jibs to avoid problems of trespass

(1) City Road project at tender stage

At tender stage the contractor needs to consider how the project will be undertaken and how long the work will take. There are many factors to think about but there is not much time to go into a lot of detail and so a risk assessment approach is often taken. This means that the estimator will identify those key factors that he thinks will influence the price and the programme, and the tender adjudication panel will decide on the extent of the risk involved and how much should be allowed in the tender to cover them.

In the case of the City Road project, several issues are of vital importance in putting together a winning bid for the contract and giving the contract team a realistic budget and programme to carry out the job and make a reasonable profit. These include:

- Site security
- Access and site layout
- Excavation and earthwork support
- Basement construction sequence
- Retaining wall demolition and reconstruction
- Choice and layout of cranage requirements
- Materials storage and handling
- Site accommodation and supervision

(1.1) City Road – pre-tender programme

At tender stage there is insufficient time to go into all the issues listed above in detail and the contractor may well have to 'take a view' in deciding his tender price. Of vital importance here is the knowledge and experience of the tender team (estimator, planner, contracts manager, etc.) and the skill and judgement of the tender adjudication panel in assessing market factors and risk issues.

At the tender stage of a project, the pre-tender programme is usually presented in bar-chart format using computer-based planning software. In the medium-sized organisation the responsibility for preparing the programme will normally be undertaken by the contracts manager, whereas the larger companies will usually have the luxury of a tender planning engineer.

Figure 10.3 illustrates a typical pre-tender programme for the City Road project, showing the main activities and the overall construction period. A realistic assessment of time periods (in weeks) for each activity is determined based on the experience of the contracts manager and planner.

Example – Steel frame and floors
Construction periods for the erection of multi-storey buildings may be based on the construction cycle time per floor which will depend on the plan size and complexity of the building layout. The overall duration of the activity will depend on how many cranes and steel erection gangs will be used and the degree of concurrency possible on the site.

Allow 2.5 weeks per floor
Overall duration $= 2.5 \times 7$ floors $=$ (say) **18 weeks**

Example – River-side retaining wall (overall wall length = approx. 150 m)
The time for the construction of the riverside retaining wall might be based on a construction cycle per 30 m length, i.e.

Form temporary access $\qquad = 1$ week
Construct wall – 5 lengths $\times 3$ weeks $= 15$ weeks
Total duration $\qquad = \textbf{16 weeks}$

An overall construction period of 70 weeks is shown for the tender stage construction period.

(1.2) City Road – site management costs

On a project of this scale, responsibility for assessing the contract preliminaries would rest on a full tender team, including the business unit or contracts manager, the planning manager and the estimator.

A team approach is essential to ensure a successful contract and decisions need to be made in respect of the main phases of work in order

PRE-TENDER PROGRAMME

City Road project

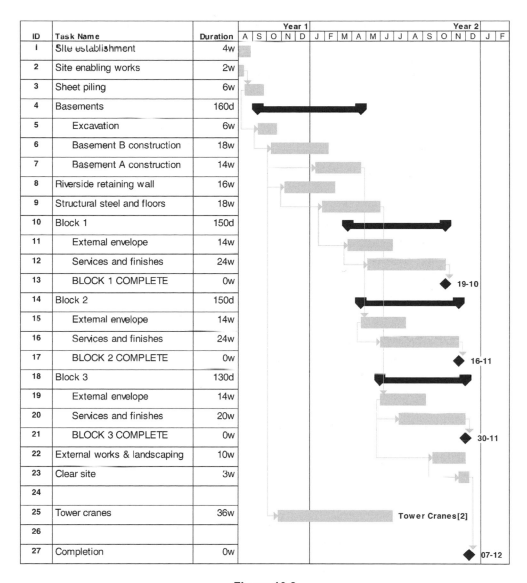

ID	Task Name	Duration
1	Site establishment	4w
2	Site enabling works	2w
3	Sheet piling	6w
4	Basements	160d
5	Excavation	6w
6	Basement B construction	18w
7	Basement A construction	14w
8	Riverside retaining wall	16w
9	Structural steel and floors	18w
10	Block 1	150d
11	External envelope	14w
12	Services and finishes	24w
13	BLOCK 1 COMPLETE	0w
14	Block 2	150d
15	External envelope	14w
16	Services and finishes	24w
17	BLOCK 2 COMPLETE	0w
18	Block 3	130d
19	External envelope	14w
20	Services and finishes	20w
21	BLOCK 3 COMPLETE	0w
22	External works & landscaping	10w
23	Clear site	3w
24		
25	Tower cranes	36w
26		
27	Completion	0w

Figure 10.3

that an appropriate organisation structure may be developed for the project.

Site staffing assessments will be required to cover engineering, quantity surveying, administration and supervision functions during the project. Company policy will dictate the extent of full-time on-site staff and which staff would only be required on site part-time.

All too often, insufficient moneys are included for site management costs at tender stage, especially when the tender adjudication panel has been looking for savings in order to make the bid more competitive. The build-up of site management costs involves consideration of:

- A realistic contract period
- The proposed site organisation structure

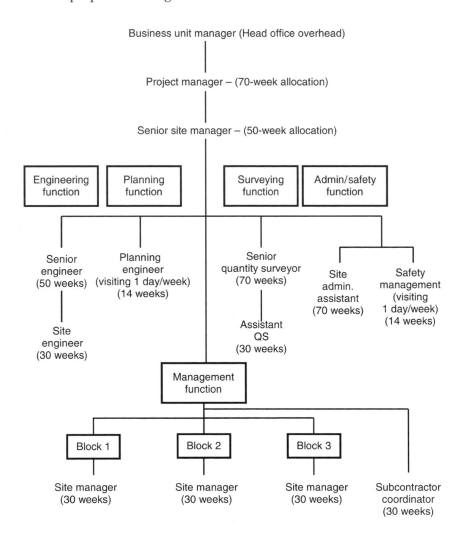

PROPOSED ORGANISATION STRUCTURE FOR PRELIMINARIES ASSESSMENT

Figure 10.4

PRELIMINARIES SCHEDULE

Site management costs

Ref.	Staff	No. off	No of weeks	Cost/weeks £	Total
A	Proj. manager	1	70	1 200	84 000
B	Snr. site manager	1	50	1 000	50 000
C	Site manager	3	90	900	81 000
D	Subcontractor/ coordinator	1	30	900	27 000
E	Senior engineer	1	50	900	45 000
F	Site engineer	1	30	800	24 000
G	Planning engineer	1	14	800	11 200
H	Senior surveyor	1	70	1 000	70 000
J	Assistant QS	1	30	800	24 000
K	Site administrator	1	70	600	42 000
L	Safety manager	1	14	800	11 200
	TOTAL Site management costs for inclusion in preliminaries				**£469 400**

Figure 10.5

- Realistic pricing of staff salaries and on-costs using a spreadsheet or a standard preliminaries build-up sheet

While there would be insufficient time to prepare a formal organogram for the project at tender stage, Figure 10.4 indicates the principles in arriving at a proposed site organisation structure for the preliminaries assessment. The time in weeks allocated to each member of the construction team is indicated.

Figure 10.5 demonstrates the use of a simple spreadsheet to assess the total site management costs inclusion in the bid. The rates used for staff costs include all employment expenses (on-costs) and company cars. A total sum of £469 400 is to be included in the contract preliminaries to cover site management costs.

(1.3) City Road – tender strategy for basement excavation

At tender stage the estimator would need to develop a strategy for the construction of the basement. This might take the form of a tender method statement.

This would consist of a simple written description of the scope of the works with the suggested construction approach for pricing purposes. Figures 10.6 and 10.7 respectively depict the basement layout and typical construction details.

Example
Scope of works:

- Basement A (80 m long × 14 m wide) is 3.5 m deep and is to be constructed alongside the new river retaining wall
- Basement B and the linked basement, (52 m long × 12 m wide) is 6.5 m deep and is to be constructed at the rear of the footpath on City Road
- Both basements are connected via a one-way ramped traffic system.
- Volume of excavation = approximately 10 400 m^3
- All excavated material must be removed via the River Place entrance/exit
- No earthwork support required along the river boundary due to the retaining wall reconstruction

Method:

- Install permanent steel sheet piles to Basements A and B using free earth support (cantilever) design
- Stage 1 excavation from ground level to a depth of −3.5 m throughout Basements A and B
- Form ramp for access to Basement A
- Install large diameter tubular steel top struts to Basement B to support sheet piles
- Stage 2 excavation to Basement B and linked basement from level −3.5 m to level −6.5 m
- Load waggons and remove excavated material from site

See Figure 10.8 for sheet piling layout plan and Figure 10.9 for details of the excavation sequence.

(1.4) City Road – cranage strategy

The allowances included in the contract preliminaries for major items of heavy plant will have a direct impact on the success of the tender bid.

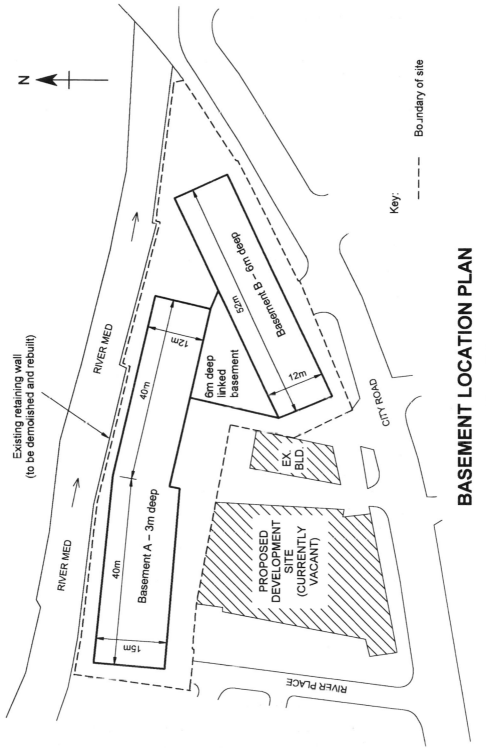

BASEMENT LOCATION PLAN

Figure 10.6

RIVER MED

RIVER MED

RIVER MED

Existing retaining wall
(to be demolished and rebuilt)

Basement A – 3m deep

40m

40m

15m

6m deep
linked
basement

12m

Basement B – 6m deep

52m

12m

CITY ROAD

RIVER PLACE

EX.
BLD.

PROPOSED
DEVELOPMENT
SITE
(CURRENTLY
VACANT)

Key:

– – – –   Boundary of site

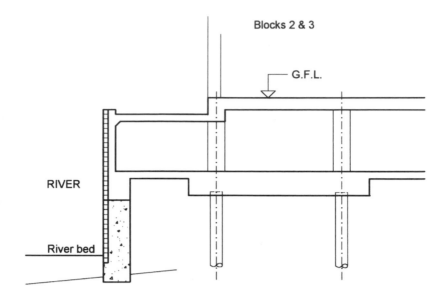

SINGLE-STOREY BASEMENT A
(Under Blocks 2 & 3)

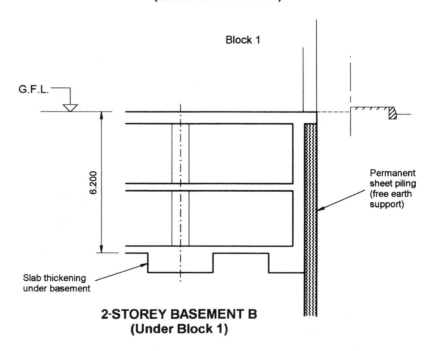

2-STOREY BASEMENT B
(Under Block 1)

BASEMENT CONSTRUCTION DETAILS

Figure 10.7

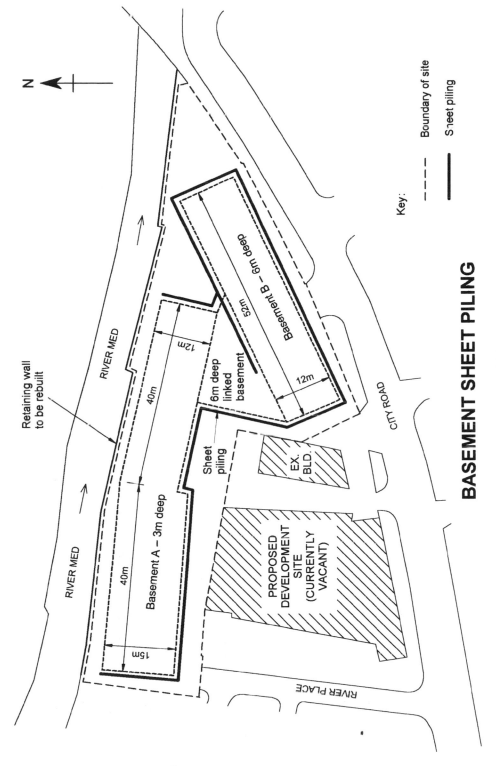

BASEMENT SHEET PILING

Figure 10.8

Key:

– – – –   Boundary of site

▬▬▬   Sheet piling

N

River Med

River Med

River Med

City Road

River Place

Retaining wall to be rebuilt

Basement A – 3m deep

Basement B – 6m deep

6m deep linked basement

Sheet piling

PROPOSED DEVELOPMENT SITE (CURRENTLY VACANT)

EX. BLD.

40m

40m

15m

12m

12m

52m

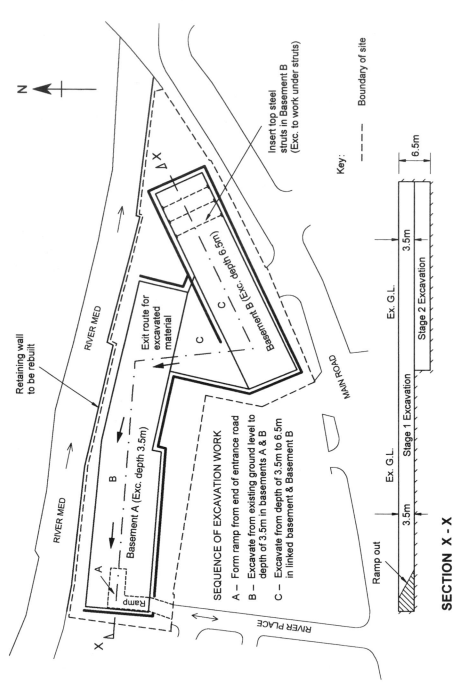

SEQUENCE OF EXCAVATION WORK

A – Form ramp from end of entrance road

B – Excavate from existing ground level to depth of 3.5m in basements A & B

C – Excavate from depth of 3.5m to 6.5m in linked basement & Basement B

SECTION X - X

SEQUENCE OF EXCAVATION WORK

Figure 10.9

Major decisions will have to be made in relation to the type and number of cranes to be included at the estimate stage. This is particularly important on sites with restricted access and limited storage space.

At the pre-tender planning stage, a number of key questions will need to be answered:

- What materials and components are being handled?
- What type and lifting capacity of crane(s) is required?
- At which stage of the project will the cranes be required and for how long?
- What facilities in the form of hardstandings, track, bases and power supplies are required?
- What size of crane would be most suitable in respect of location and reach?

The work in connection with the basement and the building superstructure poses different problems. The use of a single tower crane to serve the basement and frame may not lead to a satisfactory solution. The possibility of bringing in mobile low-pivot jib cranes for the basement construction and tower cranes for the steelwork and external envelope needs careful consideration.

Two proposals for cranage arrangements could be considered, as follows.

Proposal A
The basement construction is to be undertaken using mobile low-pivot jib cranes working from an access road formed around the basement perimeter. Figure 10.10 illustrates the crane locations relative to Basements A and B. The cranes will be used for handling reinforcement and formwork and some in situ concrete wall pours. The larger concrete pours for the foundations, basement slabs and walls will be placed using concrete pumps.

Once the basements are completed up to ground floor slab level, a tower crane will be located on each block to erect the multi-storey steel framed structure as shown in Figure 10.11.

- Duration for 2 mobile cranes = **28 weeks**
- Duration for 2 tower cranes = **18 weeks**

Proposal B
Two large 40 m radius tower cranes to be utilised to serve Basements A and B, as shown in Figure 10.11.

The tower cranes will be erected on completion of the basement excavation works and will be used for all work in connection with the basement construction and erection of the multi-storey buildings.

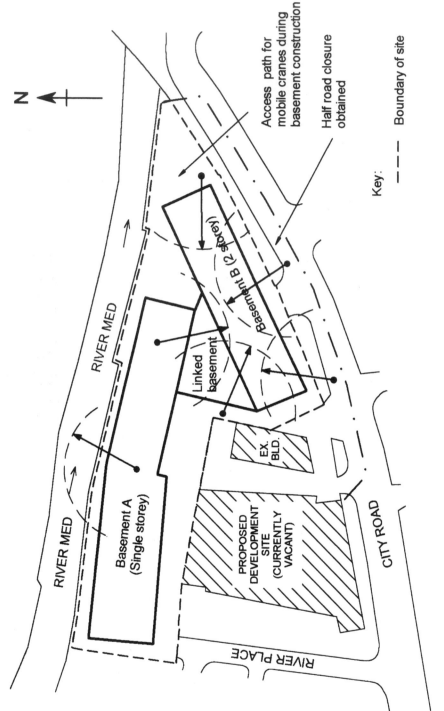

**MOBILE CRANE PROPOSALS
(for basement construction)**

Figure 10.10

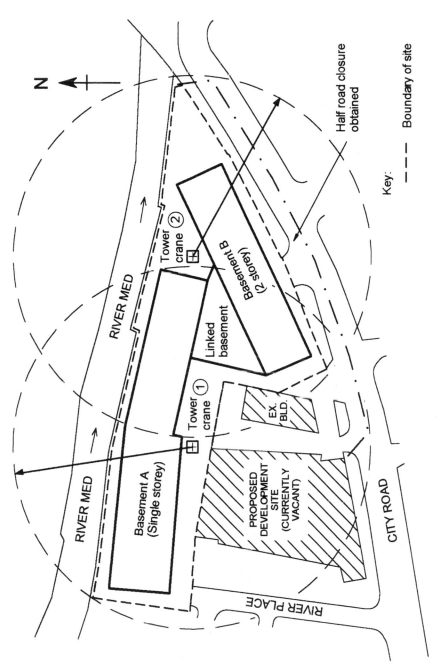

N

RIVER MED

RIVER MED

Tower crane ②

Tower crane ①

Basement B
(2 storey)

Linked basement

Basement A
(Single storey)

EX. BLD.

PROPOSED DEVELOPMENT SITE (CURRENTLY VACANT)

CITY ROAD

RIVER PLACE

Half road closure obtained

Key:

– – –   Boundary of site

TOWER CRANE PROPOSALS
(for basement and frame construction)

Figure 10.11

- Duration for 2 tower cranes = **36 weeks**

Price build-up for proposals A and B will be based on the fixed and time-related costs. The cost of providing temporary access roads for the mobile cranes in Proposal A needs to be included.

Cost studies of proposals A and B should be undertaken to establish the most economic solution on which to base the estimate.

(2) City Road project at pre-contract stage

At the pre-contract stage the contractor will know that his tender has been accepted and much more time will be invested in the planning and organisation of the project.

Detailed proposals with regard to the layout of the site need to be formulated in order to ensure that adequate space is available for site accommodation, materials storage (or laydown) areas, mortar silos, scaffolding, transport around the site and materials handling arrangements.

The pre-tender programme must now be developed into a construction (or master) programme and this needs to be supported with detailed considerations of sequence and method. The master programme will be prepared in conjunction with the contractor's project manager and a senior planning engineer based on the contract period for the project. In effect each activity on the pre-tender programme will be expanded into its constituent operations showing their dependency links in relation to each other.

Finally, it is becoming increasingly common for the contractor's tender to be judged not only on the basis of cost and programme but also in terms of quality and health and safety management proposals. This needs to be developed at the pre-contract stage into detailed arrangements to ensure that standards of quality and health and safety can be ensured on site.

The following worked examples of pre-contract proposals are given below:

- Site layout planning and materials handling
- Sequence study for basement construction
- Project quality plan

(2.1) City Road – site layout planning and materials handling

Access

- Apply for a partial road closure to City Road to provide additional space for access and site accommodation. This road would become a one-way street during the period of the contract

- Form a temporary ramp into Basement A from River Place. This allows crane and vehicular access to all basement areas and to the adjoining riverside retaining wall construction
- Five access and storage areas have been indicated on the site layout plan in Figure 10.12. These are designated as areas A, B, C, D and E

Location of site accommodation

- Two-tiered fully serviced site accommodation is to be located on City Road, as shown on the site layout plan. Two access gates are to be provided in the hoarding to allow access to Area A and Area D
- Basement A is sufficiently wide to allow subcontractors' cabins to be located here for a temporary period until the installation of the basement piles and floor slab
- Following erection of the steel frame to Blocks 1, 2 and 3, site accommodation for subcontractors will be provided in the ground floor areas of each block

Materials storage areas

- Tower cranes A and B are to be erected in the locations indicated on the layout plan. Area A on City Road has been designated an unloading area for main materials as vehicles can reverse into this area and be unloaded by tower crane B. Once Basement A is completed, the area marked E can be used for unloading materials with tower crane A. This area will no longer be required for access
- Arrangements for materials storage to be as follows:
  Area A   Main vehicle unloading area for key materials
  Area B   Five metre wide area in front of blocks 2 and 3 to be used for palletted brick storage
  Area C   Ground floor storage area part over linked basement
  Area D   Key storage area with access to City Road via gates in the hoarding. Site silo for mortar mixing and supply to be located in this area
  Area E   Unloading area
- Due to lack of storage space, phased deliveries for the following key materials is to be considered:
  o   Bricks and blocks
  o   Stud partitioning
  o   Internal joinery
  o   Heating and ventilation equipment
- Consideration might also be given to the pre-loading of the floors of the building during frame erection with pre-packaged materials and components boxed-up individually for each apartment type:
  o   Door linings
  o   Pre-cut architraves

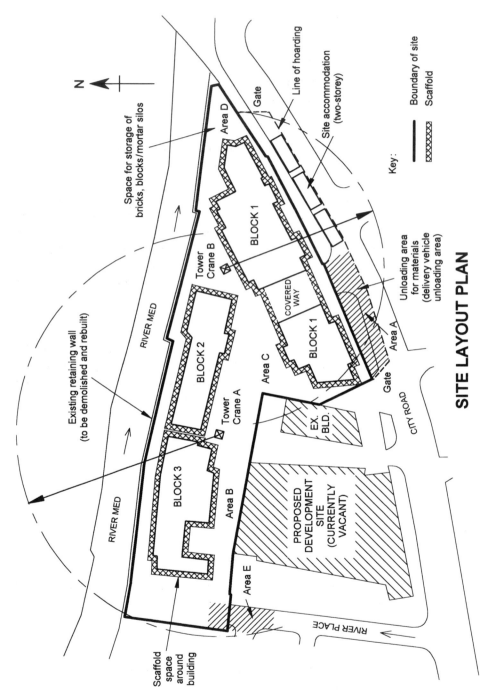

SITE LAYOUT PLAN

Figure 10.12

- o Ironmongery
- o Electrical wiring sets
- o Heating packs

(2.2) City Road – sequence study for basement construction

The following diagrammatic sequence study has been developed for the construction of Basement A.

Description of works
The work involves the construction of a single basement 80 m × 15 m × 3.5 m deep comprising reinforced concrete floor and thickening beams, associated piling works and reinforced concrete walls.

The basement is to be permanently sheet piled on three sides based on the free earth support method (cantilever). The long side of the basement is adjacent to the new riverside retaining wall. See Figures 10.6 and 10.7 for details of the basement layout and typical cross-sections.

The sequence study is to include all operations to the underside of the ground floor slab level. The suspended in situ concrete waffle slab forming the ground floor is not included.

Development of construction sequence
The planner or construction manager needs to carefully consider all the stages involved in the construction sequence, including intermediate activities and any temporary works required. This can only be gained from experience on contracts – it cannot be learnt from a book. The following is offered as a guide to the thought process:

- Consider the work sequence or stages involved:
  - o excavation work
  - o forming the basement slab
  - o constructing the basement walls
  - o links between fixing reinforcement
  - o erecting formwork, etc.
- Consider overlapping operations
- Think about start to start relationships:
  - o When can the floor construction commence in relation to the excavation works?
  - o How will the floor be poured? How many bays?
  - o When can the basement walls start relative to the floor construction?
- Consider continuity of work for operative gangs:
  - o Steelfixers
  - o Carpenters
  - o General labourers for concreting

- Think about use of formwork (shuttering):
  - How many lifts?
  - Is formwork to be full height?
  - How many uses for wall shutters?
- Make allowance for the striking times of formwork:
  - Beam sides and vertical wall faces – next day
  - Soffites to beams and suspended floor slabs – 7–10 or 14 days depending on specification and possible evidence of concrete cube strengths
- Consider utilisation of plant resources:
  - Handling reinforcement, formwork and concrete
  - Concrete will normally be pumped with maybe three or four pours arranged for the day when the concrete pump is on site

Sequence of operations

- Stage 1   Install permanent sheet piles (three sides of excavation)
- Stage 2   Excavate basement and form piling mat
- Stage 3   Piling to basement floor
- Stage 4   Excavate thickening beams to basement floor
- Stage 5   Cut off piles and blind excavation with concrete
- Stage 6   Fix steel reinforcement to basement floor slab and thickenings
- Stage 7   Concrete basement slab
- Stage 8   Fix formwork and reinforcement to basement walls
- Stage 9   Concrete basement walls
- Stage 10   Strip formwork to basement walls

Figures 10.13–10.17 show a sequence study of the basement construction stages described above for Basement A. Figure 10.13 shows the basement plan and cross-sections and, in particular, the in situ concrete floor slab thickening details. The sequence study has assumed that the thickenings and floor slab will be poured at the same time irrespective of whether Proposal A or B is chosen for the floor pour sequence (Figure 10.17 refers). As an alternative, the contractor could pour the thickenings up to the underside of the floor slab and then pour the floor as a subsequent operation.

(2.3) City Road – project quality plan

Upon notification of the award of the contract, the construction manager will initiate the preparation of the quality plan. This will need to be completed before work commences on site. It should be noted that the following is a basic quality plan and some clients often require a more comprehensive document linked into the company quality procedure.

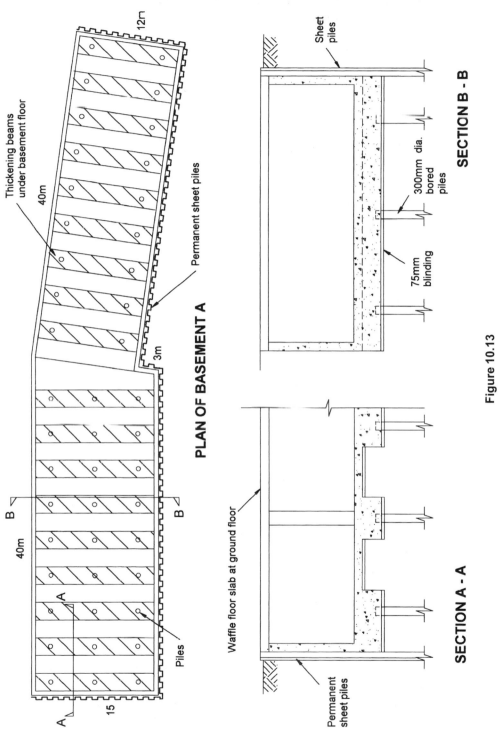

Thickening beams under basement floor

Permanent sheet piles

40m

3m

12n

PLAN OF BASEMENT A

40m

B

B

A

A

15

Piles

Sheet piles

300mm dia. bored piles

75mm blinding

SECTION B - B

Waffle floor slab at ground floor

Permanent sheet piles

SECTION A - A

Figure 10.13

DIAGRAMATIC CONSTRUCTION SEQUENCE

STAGE 1 – Insert permanent sheet piles

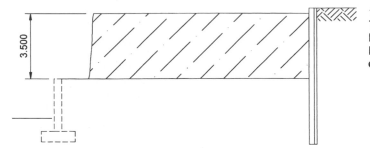

110 linear metres of
7–8 m long piles to
be driven. Piles
left in as rear face
of basement wall

STAGE 2 – Excavate basement to depth of 3.5m

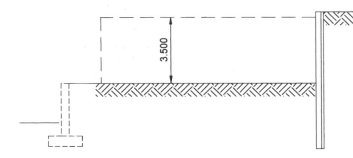

Excavation volume
approx. 3800 m³
to be excavated
and carted to
tip-off site

Duration = 3800
          ─────────
          300 m³/day

Approx. = 12 days

STAGE 3 – Piling to basement slab

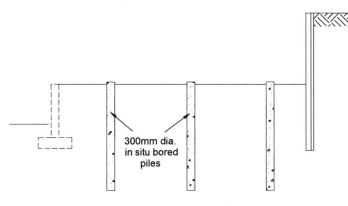

300mm dia.
in situ bored
piles

Approx. 40 - 300mm dia.
bored piles in basement
Allow 5 days set up/
dismantle equipment

8 piles/day - 5 days
drilling

Approx. = 10 days overall

Figure 10.14

DIAGRAMATIC CONSTRUCTION SEQUENCE

STAGES 4 & 5
Excavate thickening beams to basement
floor – Cut off piles and blind slab

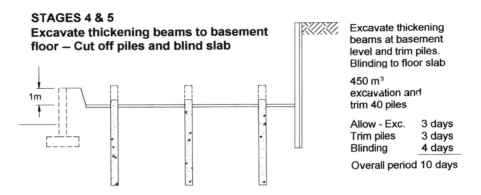

Excavate thickening
beams at basement
level and trim piles.
Blinding to floor slab

450 m³
excavation and
trim 40 piles

Allow - Exc.   3 days
Trim piles   3 days
Blinding   4 days

Overall period 10 days

STAGE 6 – Reinforcement to floor slab and thickenings

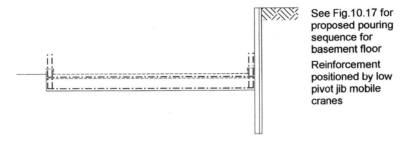

See Fig.10.17 for
proposed pouring
sequence for
basement floor

Reinforcement
positioned by low
pivot jib mobile
cranes

STAGE 7 – Concrete basement slab and thickenings

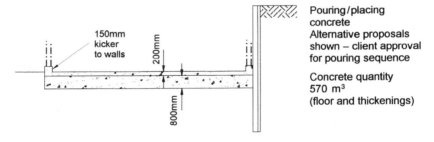

150mm
kicker
to walls

200mm

800mm

Pouring/placing
concrete
Alternative proposals
shown – client approval
for pouring sequence

Concrete quantity
570 m³
(floor and thickenings)

Note – Thickenings may be concreted as a separate pour, prior to basement slab

Figure 10.15

Contents of the quality plan
The contract-specific quality plan is based on ISO 9002 and the company's procedures manual and contains the following headings:

- Quality statement
- Contract particulars

DIAGRAMATIC CONSTRUCTION SEQUENCE

STAGE 8
Reinforcement and formwork to walls
using full height formwork

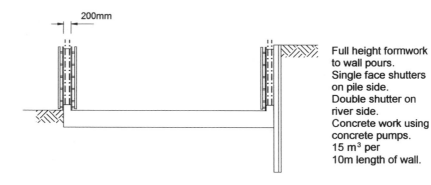

Full height formwork
to wall pours.
Single face shutters
on pile side.
Double shutter on
river side.
Concrete work using
concrete pumps.
15 m^3 per
10m length of wall.

STAGE 9 – Concrete to walls

Mobile crane to be used for handling reinforcement and formwork panels.
Max. length of pour 10 m on plan.
Reinforcement to precede formwork as continuous operation around basement.

Concrete pours to alternate floor bays may be coordinated with wall pours for
economy of concrete pumping.

Sequence of constructing the wall and floor as shown on floor plan.
Sequence of wall pour, Fig.10.17 refers, to follow flooring sequence chosen.

Figure 10.16

- Organisation structure for the project
- Specific staff responsibilities
- Specialist services
- Control of workmanship and method statements
- Issue of drawings
- Control of subcontractors
- Control of on-site materials

Quality statement
'The company aims to meet the requirements of customers and of the
industry and recognises that quality is a constantly changing practice.
The company recognises that quality is everybody's responsibility and
it is committed to training and establishing an organisation that will
ensure a total understanding by all employees.

To achieve its requirements, the company has adopted a policy to
operate a coordinated quality system which conforms to the requirements

Considerations for concreting basement slab

Proposal A

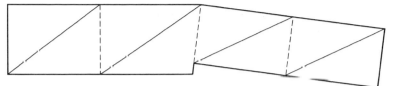

Pour floor slab in 4 pours
Approx. 140 m³ per pour
or in 8 pours at 70 m³ per pour

Proposal B

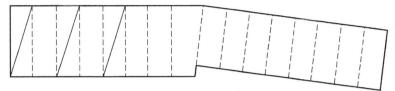

Pour slab and thickenings in 5 m wide alternative
bay construction – approx. 36 m³ per pour
3 bays may be constructed per concrete pump visit – 108 m³/3 days

All floor pours to be concrete pumped from a location inside the basement area
or with the pump located on storage area B

Wall pour sequence

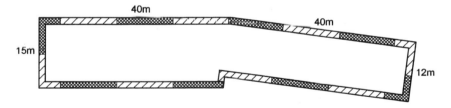

FLOOR AND WALL POUR SEQUENCE

Figure 10.17

of BS 7850 and ISO 9002. Mr Baker, Regional Director, has been appointed with the responsibility and authority to manage the adopted quality management system.'

Contract particulars
This part of the quality plan contains a summary of the conditions of contract, making reference to specific clauses in the form of contract. A description of works together with the contract commencement and completion date should be given.

A summary should also be given of the main participants involved in the contract, i.e. client, architect, design team and employer's representative. A list of the contractor's personnel engaged in managing the project should be included, indicating individual responsibility for construction management, surveying and health and safety.

Organisation structure for project
A diagram showing the organisation structure for the project is commonly provided. This might be a 'family tree' diagram or perhaps a linear responsibility matrix showing roles and responsibilities, as suggested by Walker (2002).

Figure 10.18 indicates the line management structure from construction director through to site level with emphasis on safety management.

Specific responsibilities
This part of the quality plan contains a schedule of responsibilities allocated to the construction project manager, assistant site manager, site engineer and assistant surveyors and foreman. Responsibilities may be defined under the following general headings:

- Pre-start arrangements
- General site administration
- Management reports
- Materials management
- Site surveying functions

Specialist services
This section contains a check-list for the design of temporary works, formwork and the permanent works.

The parties responsible for the design of various elements of the permanent and temporary works will be identified, including the novated design team, works package contractors and suppliers of specialist components and equipment.

Control of workmanship and method statements
This section highlights operations from a specific list of activities which relate directly to the project. The activities are judged to be quality critical to the success of the project. In principle they relate to the company's quality assurance scheme, to method statements affecting construction operations and to health and safety.

A schedule of method statements already prepared for early site operations and following operations would be provided.

Issue of drawings
This section contains instructions for the issue and distribution of

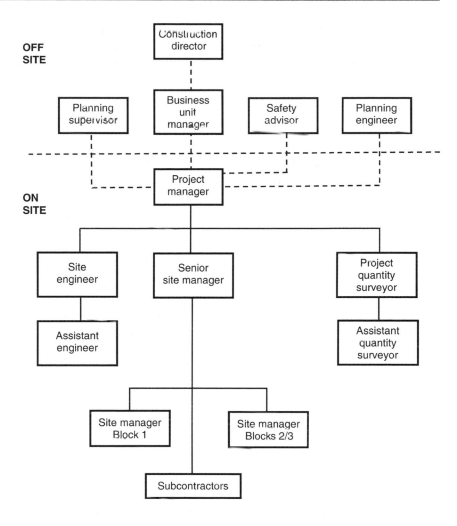

**CITY ROAD PROJECT
ORGANISATION STRUCTURE**

Figure 10.18

drawings to all parties, i.e. head office, site and design team. The quick and efficient communication of information is vital to avoid delays and disruption on site.

Control of subcontractors
A schedule of operations relating to work undertaken by subcontractors is listed in this section. Their responsibility for developing project-specific method statements and attendance at meetings is outlined.

Control of on-site materials

This is a schedule of tests to be undertaken on site materials and relates to quality checks and destructive and non-destructive materials testing (i.e. cube tests for concrete, sonic tests for piles, etc.).

(3) City Road project at contract stage

(3.1) City Road – subcontract procurement and management procedures

The main construction operations on site have been divided into a number of individual subcontract work packages. A number of the packages involve some degree of design responsibility. The subcontract packages include:

- Sheet piling (earth support)
- Formwork, reinforcement and concreting of the basement works
- Structural steelwork, design and erect
- Metal deck floors
- External brickwork and blockwork
- Heating, ventilation and domestic services
- Lift installation

In this part of the case study, subcontract management arrangements are outlined taking a typical subcontract package through from the pre-contract stage of the contract to the end of contract review. The sheet piling subcontract is used as the example in relation to subcontract procedures at the pre-contract and contract stage.

Tender procedures

At the tender stage of the project enquiries have been sent to three sheet piling contractors, and design proposals and firm prices have been received. Proposals from R&M Piling have been included in the tender submission.

Pre-contract procedures

These include:

- Approval of subcontract list
- Preparation of subcontract procurement schedule
- Review procurement schedule
- Subcontract pre-selection process
- Subcontract pre-placement meeting
- Place subcontract order (QS function) with R&M Piling
- Notification to commence on site

Contract procedures
These include:

- Start on site including
  - o Submission of risk assessments and method statements
  - o Submission of insurance requirements
  - o Signed order
- Site induction procedures
- Quality inspection plan
- Agree subcontract target programme
- Subcontract weekly and monthly site meeting involvement in short-term planning
- Procedures at site management level
- Agreeing any subcontract variations and extensions of time
- Dealing with subcontract payments and final account
- Subcontract notice of completion
- End of contract review of subcontractors' performance

Subcontract management is a team effort between the project manager, the senior quantity surveyor and the planning engineer. Management of the subcontractor at the site face is undertaken by the appropriate site management personnel.

The above procedures are those implemented by the main contractor on the City Road project. This is a good example of a contractor/subcontractor relationship aimed at achieving a successful contract for all parties involved in the contract.

(3.2) City Road – procedures manual for planning and control

The following extracts are taken from the procedures manual of a large contracting organisation. They have been simplified and applied to the City Road project.

Role of the planning department
To perform the following functions:

- Initial programming for a project at the client design stage of a design and build project enquiry
- Detailed tender programme preparation
- Contract master programme preparation
- Preparation of a procurement programme and procurement schedules
- Progress evaluation at site level (including the preparation of an independent progress report on a monthly basis)
- Procurement updating

- Review of information flow
- Preparation of an as-built programme as work proceeds

Responsibilities of the site planning engineer
The planner will carry out the following tasks:

- Review project documentation
- Review the project drawings
- Review procurement and contractual considerations
- Review the proposed phasing and sequencing of work
- Undertake direct comparisons of proposed works against actual similar work previously completed
- Undertake a review of construction methods with the site manager
- Prepare the 2-weekly short-term programmes in conjunction with the appropriate site personnel
- Prepare a weekly and monthly report on progress for submission to the site manager and project manager with a copy to the chief planning engineer

Contract control documentation
On award of the contract, the following control documents form the basis of the link between the planner and the site construction team:

- The master contract programme
- The target programme
- The materials requirement schedules
- The subcontract procurement schedules
- Information requirement schedules

Operating procedures when preparing contract programmes
The purpose of these procedures is to produce a master contract programme to be issued to the design team and client and to provide a construction target programme for construction and monitoring purposes. Figure 10.19 illustrates the procedures to be followed.

(3.3) City Road progress records and reports

Liaison between site-based planning engineer and site managers

- The site-based planning engineer will be responsible for following the company procedures as indicated in the procedures manual, as outlined above
- Progress on the target programme will be recorded using a computer software package at both weekly and monthly intervals

PREPARATION OF MASTER CONTRACT PROGRAMME

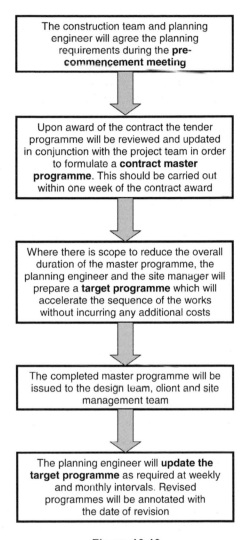

The construction team and planning engineer will agree the planning requirements during the **pre-commencement meeting**

Upon award of the contract the tender programme will be reviewed and updated in conjunction with the project team in order to formulate a **contract master programme**. This should be carried out within one week of the contract award

Where there is scope to reduce the overall duration of the master programme, the planning engineer and the site manager will prepare a **target programme** which will accelerate the sequence of the works without incurring any additional costs

The completed master programme will be issued to the design team, client and site management team

The planning engineer will **update the target programme** as required at weekly and monthly intervals. Revised programmes will be annotated with the date of revision

Figure 10.19

- The site-based planning engineer will work closely with the appropriate site manager in the preparation of detailed stage programmes in relation to major sections of the work. During the undertaking of the works on site, short-term two-weekly programmes will be prepared and implemented by both the planner and site manager in charge of that particular work section

MONTHLY PROGRESS REPORT

Contract	City Road	Contract No	1714
Report No	5	Name	A Smith
Date	16 Feb	Contract Period	70
Weeks completed	20	Weeks remaining	50
Contract Completion Date		22 May	
Target Completion Date		4 Feb	
Anticipated completion		4 Feb	

MAIN PHASES OF WORK	PROGRESS COMMENTARY
Site establishment	Commenced on time, now 100% complete
Sheet piling	Started on time, completed 2 weeks later than planned due to work in connection with the adjacent river retaining wall. Work now 100% complete
Riverside retaining wall	Started 4 weeks later than planned due to variable ground conditions and changes in the foundation design. Currently 5 weeks behind programme. Work now 60% complete
Basement A	Excavation works completed on time. Delays to the wall construction of Basement B have delayed commencement of the basement slab construction. Basement works are currently 4 weeks behind programme. Work now 20% complete
Basement B	Basement concrete and formwork started 3 weeks late due to design delays. Work currently progressing to basement walls at second level. Basement works currently 4 weeks behind programme. Work currently 80% complete
PROCUREMENT	
Subcontracts	Orders placed for steelwork which is due for erection on Block 1 at week 26. The commencement may have to be delayed by 3 weeks
Materials	Brick samples approved and orders placed

COPIES TO	Site manager ☐	Project manager ☐	Chief planning eng ☐

Figure 10.20

Monthly Progress Reports

Fig 10.20 contains an extract from a monthly site report produced at the end of week 20 of the contract. This is produced for the project review meeting at the end of each month. It is compared with the independent report produced by the head office planner. The report is presented in operational format, supported by an as-built bar chart together with the updated target programme.

10.4 Case study No. 2 – Shopping mall project

Project brief

> The project involves the construction of a £300 million shopping mall complex on the outskirts of Manchester. The contract has been awarded on the basis of a management contract arrangement. The management contractor was appointed to assist with the project feasibility study and to provide advice on buildability at the design stage.
>
> The value of the building shell and core is approximately £160 million and this forms the bulk of the work contained in the management contract. The shell and core involves the coordination of some 60 work packages.
>
> The overall contract period, including the clients fitting out, has been assessed at 32 months. The value of the fitting out work is in the order of £100 million.

Worked examples

(1) Project relationships

Figure 10.21 indicates the relationship between the client/client's project team/the design team and the management contractor. The client's project team consists of eight members of the client's commercial team who will be ultimately responsible for taking over the complex on completion. A team of representatives from the design team is based on site in order to coordinate information requirements.

Figure 10.22 illustrates the management contractor's site team established to manage the project. The site management staff approximates to some 30 clerical and management personnel. The team is under the control of a site-based project director assisted by a construction executive responsible for the management of site activities. The management of the project is split up into six areas (see Figure 10.22):

- Financial planning
- Package
- Construction
- Procurement
- Planning
- Design

A consulting land survey company has also been engaged to provide the main grid references for setting out the foundation works and car park areas. Figure 10.23 illustrates a plan of the project indicating the proposed nine work zones. This relates to the sequence of work for the

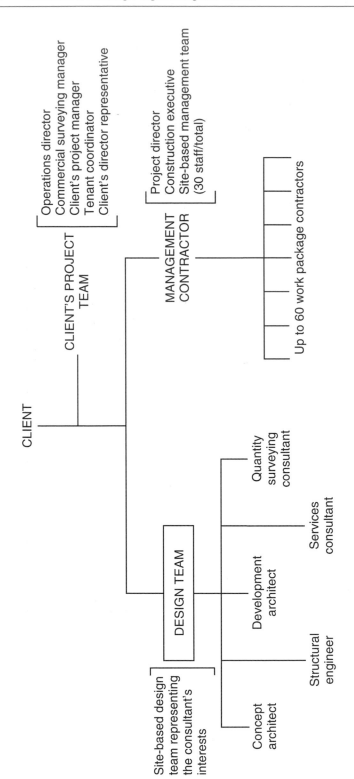

CLIENT

CLIENT'S PROJECT TEAM
- Operations director
- Commercial surveying manager
- Client's project manager
- Tenant coordinator
- Client's director representative

MANAGEMENT CONTRACTOR
- Project director
- Construction executive
- Site-based management team
- (30 staff/total)

Up to 60 work package contractors

DESIGN TEAM

Site-based design team representing the consultant's interests

- Concept architect
- Structural engineer
- Development architect
- Services consultant
- Quantity surveying consultant

RELATIONSHIP BETWEEN CLIENT/DESIGN TEAM AND MANAGEMENT CONTRACTOR

Figure 10.21

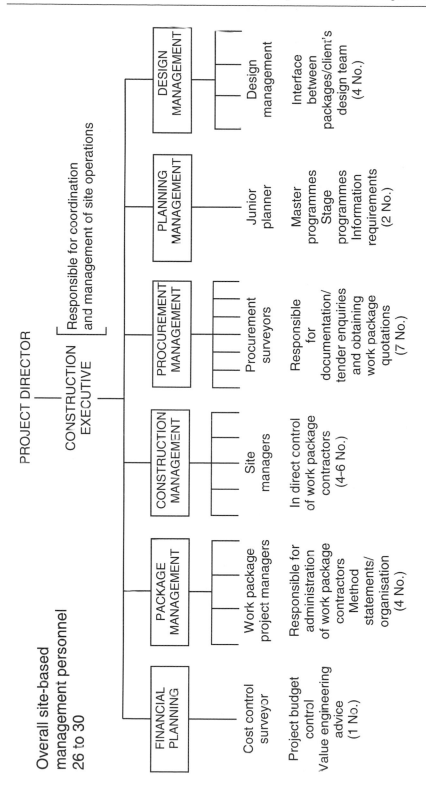

PROJECT DIRECTOR

CONSTRUCTION EXECUTIVE — Responsible for coordination and management of site operations

Overall site-based management personnel 26 to 30

FINANCIAL PLANNING

Cost control surveyor

Project budget control
Value engineering advice
(1 No.)

PACKAGE MANAGEMENT

Work package project managers

Responsible for administration of work package contractors
Method statements/ organisation
(4 No.)

CONSTRUCTION MANAGEMENT

Site managers

In direct control of work package contractors
(4–6 No.)

PROCUREMENT MANAGEMENT

Procurement surveyors

Responsible for documentation/ tender enquiries and obtaining work package quotations
(7 No.)

PLANNING MANAGEMENT

Junior planner

Master programmes
Stage programmes
Information requirements
(2 No.)

DESIGN MANAGEMENT

Design management

Interface between packages/client's design team
(4 No.)

MANAGEMENT CONTRACTOR'S SITE-BASED TEAM

Figure 10.22

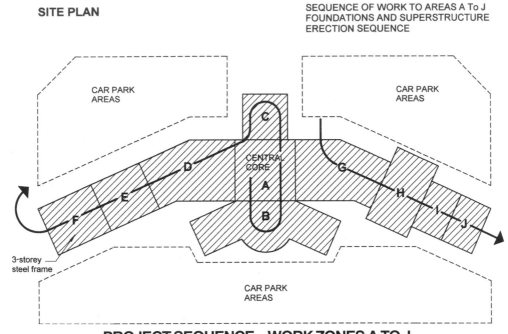

SITE PLAN

SEQUENCE OF WORK TO AREAS A To J
FOUNDATIONS AND SUPERSTRUCTURE
ERECTION SEQUENCE

CAR PARK
AREAS

CAR PARK
AREAS

C

CENTRAL
CORE

D

A

G

E

H

F

B

I

J

3-storey
steel frame

CAR PARK
AREAS

PROJECT SEQUENCE – WORK ZONES A TO J

Figure 10.23

piling, substructure, frame erection and suspended floors. Joint consultations between the design team and management contractor have resulted in the development of the construction sequence A to J.

(2) Project planning and control strategy

Figure 10.24 shows the master programme with the monetary values of the main work activities allocated to time. The cumulative value forecast has been plotted across the bar chart. This indicates valuations in the order of some £9 million per month during peak contract activity.

Computers are used extensively on the project in order to handle the vast amount of information passing between the design team and management contractor. Power Project (Asta Development) is being used for the coordination of the work package programmes. Programmes submitted by the work package subcontractors are transferred to Open Plan (project management software) in order to link them into the main contract network programme.

The responsibility for planning the project lies with a senior planner and an assistant. The senior planner's responsibilities include:

- Preparation of the master programme
- Coordination of the work package master programme

Duration – 32 months
Project value – £160M

MASTER PROGRAMME

SHOPPING PRECINCT

Month No.

WORK AREAS	Dur	Val	1	2	3	4	5	6	7	8	9	10	11	12	13	14	15	16	17	18	19	20	21	22	23	24	25	26	27	28	29	30	31	32
Enabling work	4	2	0.5	0.5	0.5	0.5																												
Substructure	8	4					0.5	0.5	0.5	0.5	0.5	0.5	0.5	0.5																				
Structural frame	8	16							2	2	2	2	2	2	2	2																		
Fire protection	10	5									0.5	0.5	0.5	0.5	0.5	0.5	0.5	0.5	0.5	0.5														
External envelope	12	25											2	2	2	2	3	2	2	2	2	2	2	2										
Mec./elec. services	18	54							3	3	3	3	3	3	3	3	3	3	3	3	3	3	3	3	3	3								
Mall finishes	12	24																			2	2	2	2	2	2	2	2	2	2	2	2		
Ext. wks per road	6	6					1	1	1																							1	1	1
Ext. car parks	24	24									1	1	1	1	1	1	1	1	1	1	1	1	1	1	1	1	1	1	1	1	1	1	1	1
Total value		160																																
Value/month			0.5	0.5	0.5	0.5	1.5	1.5	6.5	5.5	7	7	9	9	8.5	8.5	7.5	6.5	6.5	6.5	8	8	8	8	6	6	3	3	3	3	3	4	2	2
Cumulative value			0.5	1	1.5	2	3.5	5	11.5	17	24	31	40	49	57.5	66	73.5	80	86.5	93	101	109	117	125	131	137	140	143	146	149	152	156	158	160

Cumulative value curve

Figure 10.24

- Preparation of the stage programme for each work package indicating key dates for the start and finish of each of the nine work zones
- Coordination of design team information requirements
- Monitoring contract progress with the package managers
- Attending project coordination meetings with the project team and client's representatives
- Attending work package contractor interviews at the package negotiation stage

Each of the work package contractors is responsible for short-term planning within each of their work zones. This is based on 4–6 weekly short-term planning programmes to be agreed with the package managers.

Figure 10.25 indicates the relationships between the work package master programme, stage programme and 6-weekly short-term programme as applied to the piling and substructure work package. It is becoming the norm on major projects for subcontractors to be responsible for the planning of their own work, and they must be willing to include for this risk at the tender stage.

During the project a series of project site meetings are held in order to report on the project progress position both weekly and monthly. Meetings are essential in order to review information requirements between the client's project team, the design team and the management contractor.

10.5 Case study No. 3 – Housing project

Project brief

The project consists of the construction of 2 eight-storey reinforced concrete framed apartment blocks, which are to be externally clad in brickwork, and 28 low-rise housing units. The foundations to each block are to be piled and incorporate a raft foundation. The balance of the development consists of the erection of 28 low-rise units contained in six terraced blocks, as shown on the plan in Figure 10.26. The adjacent site roadway and site service connections have been completed as part of a separate contract.

The project is £3.5 million in value and the buildings are to be handed over in phases to an agreed schedule.

Worked examples

(1) Construction strategy
(2) Pre-tender method statement based on the agreed strategy
(3) Pre-tender programme based on the method statement

PROGRAMMING STAGES DURING CONSTRUCTION

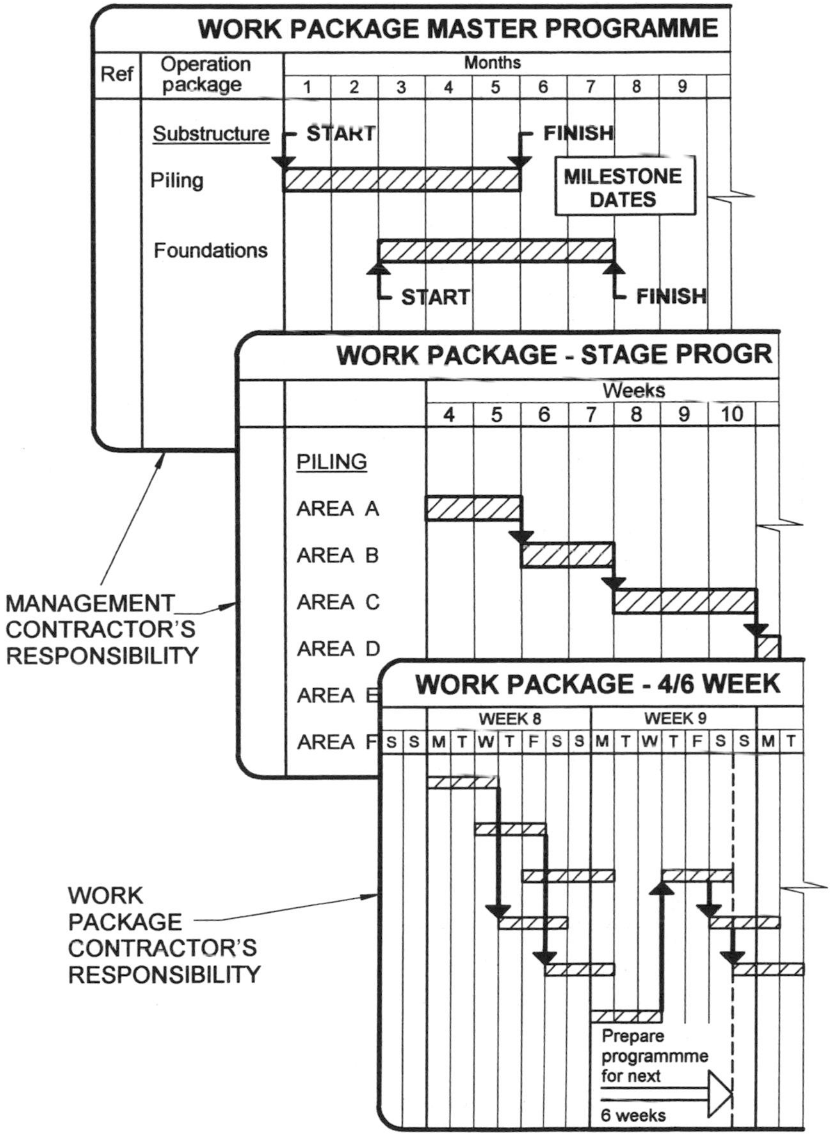

Figure 10.25

(1) Construction strategy

Consideration could be given to dividing the site into two distinct sections and managing the project effectively as two separate contracts (albeit under one project manager). Section 1 would involve the construction of the two tower blocks and Section 2 the construction of the low-rise dwelling units.

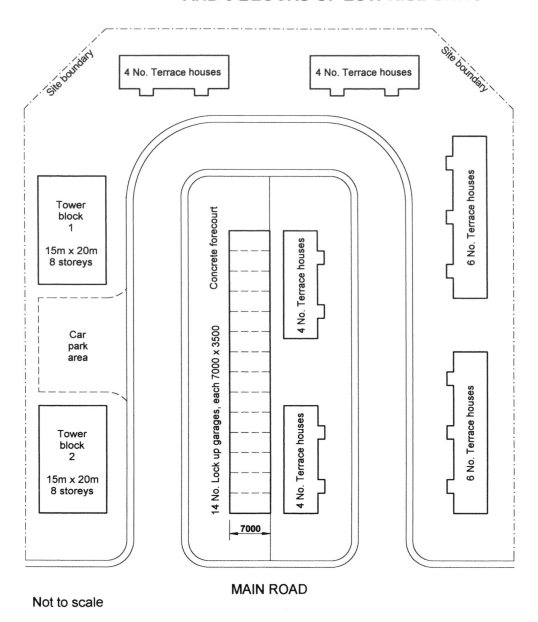

SITE PLAN   **TWO 8-STOREY APARTMENT BLOCKS AND 6 BLOCKS OF LOW-RISE UNITS**

MAIN ROAD

Not to scale

Figure 10.26

In this way both Sections 1 and 2 could be started at the same time, so saving in time-related costs. Alternatively, the construction of the tower blocks could start prior to the low-rise units. A fast track approach could be considered on the project, as described in option 2 below.

There are two possible options for Section 1:

- **Option 1** – Construct tower block A and then move the tower crane to construct tower block B. The labour gangs would likewise move from block to block as the work progressed
- **Option 2** – Construct tower blocks A and B simultaneously. This will involve erecting a separate tower crane on each of the blocks. Likewise, separate labour gangs would be required for each block

The work in connection with the piling of both tower blocks would be undertaken during one visit to the site. Consideration could be given to using two piling gangs in order to facilitate piling of both tower blocks at the same time.

A senior management decision will be necessary in order to agree the project strategy at this stage, as this will directly affect the contractor's bid in respect of the overall project period.

(2) Pre-tender method statement

Figure 10.27 illustrates a pre-tender method statement based on the decision to undertake a fast track approach as outlined in option 2. Figure 10.28 indicates the sequence of work illustrated on the layout plan.

(3) Pre-tender programme

- Assessment of duration for each tower block:
  Foundations and piling work   =   4 weeks
  Reinforced concrete frame:
  allow a 2-week cycle per floor
  = 8 floors × 2 weeks/floor   =   16 weeks
  External brickwork cladding   =   4 weeks
  Internal finishes   =   8 weeks
  Total   =   32 weeks (**8 months**)
- Assessment of duration for low-rise blocks:
  Four-unit block   =   16 weeks (**4 months**)
  Six-unit block   =   20 weeks (**5 months**)

Figure 10.29 shows the pre-tender programme based on constructing the tower blocks A and B separately (option 1).

The pre-tender programme in Figure 10.30 illustrates a fast track approach to the project based on the pre-tender method statement for constructing the tower blocks simultaneously (option 2). The start to start delay of 2 months between blocks 1 and 2 is to allow for continuity of work for the piling contractor.

MULTI-STOREY APARTMENTS AND LOW-RISE UNITS	Sheet 1

PRE-TENDER METHOD STATEMENT

1 – SITE ESTABLISHMENT

Loop access road constructed prior to commencement of work.

Site compound to be located in area of block 6, with entry and exits to adjacent road.

Fully serviced compound to be maintained throughout the contract period until commencement of work to block 6.

At this stage the compound may be reduced in size to accommodate the rear garage area.

Garages at the rear of block 5 to be constructed at early stage in project as they are to be used as covered storage.

2 – TOWER BLOCKS 1 AND 2	SUBSTRUCTURE WORK

Piling to blocks 1 and 2 to be completed in single visit to site.

Tower blocks to be constructed using separate gangs on substructure and superstructure work.

Concrete pumping to be utilised for foundation beams and ground floor slab construction.

3 – TOWER BLOCKS 1 AND 2	SUPERSTRUCTURE WORK

Fast track approach using a separate tower crane located in front of each block.

Construction sequence of 2 weeks per floor to be achieved with aid of patent formwork system and complete set of formwork soffits for two floors of the building.

Floor pours to be concrete pumped.

Figure 10.27a

10.6 Case study No. 4 – Canal project

General description

The project consists of the refurbishment of a disused canal and lock as part of a project to create a marina facility for pleasure boats and yachts. The contract value is £1 million and the conditions of contract are the ICE Conditions. General arrangement drawings of the existing site and details of the proposed works are given in Figures 10.31 and 10.32.

MULTI-STOREY APARTMENTS AND LOW-RISE UNITS	Sheet 2

PRE-TENDER METHOD STATEMENT

4 – LOW-RISE UNITS	SUBSTRUCTURE WORK

Substructure works to low rise units to commence at week 1 of contract.

Low-rise units to be managed by separate construction manager.

Foundation gangs to be moved from block to block in sequence shown on layout drawing.

Four week construction cycle per block to be developed for foundations and services to ground floor level.

5 – LOW-RISE UNITS	SUPERSTRUCTURE WORK

A 16 and 20 week construction cycle per block to be developed with phased handover as illustrated on pre-tender programme.

6 – LOW-RISE UNITS	BUILDING SERVICES

Full continuity of specialist services work to be achieved for subcontract service trades (electrical, heating and plumbing services).

7 – PROJECT PROGRAMME

A fast track approach to the project has been taken with an overall completion date of 10 months.

Phased handover programme to be prepared and submitted for client's approval.

Figure 10.27b

Project details

The work involves cleaning out the existing canal and lock together with new work comprising:

- a precast concrete retaining wall with in situ concrete capping
- an in situ concrete retaining wall with gabion wall behind
- guniting to the existing lock walls
- drainage works

SITE PLAN   TWO 8-STOREY APARTMENT BLOCKS AND 6 BLOCKS OF LOW-RISE UNITS

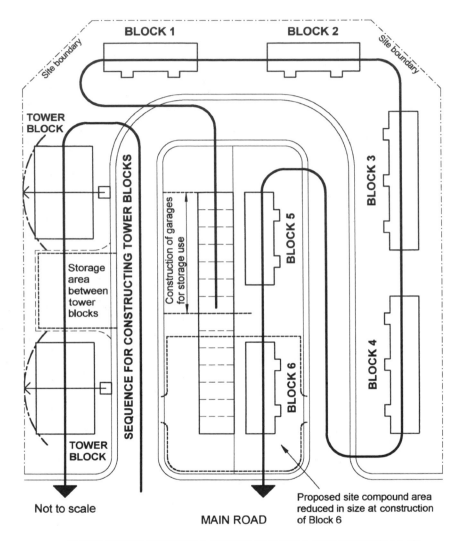

CONSTRUCTION SEQUENCE PROPOSALS

Figure 10.28

- a small pumphouse comprising loadbearing brickwork and a concrete flat roof
- foundations for a new swingbridge
- refurbishment of two pairs of existing lock gates
- associated earthworks and temporary works

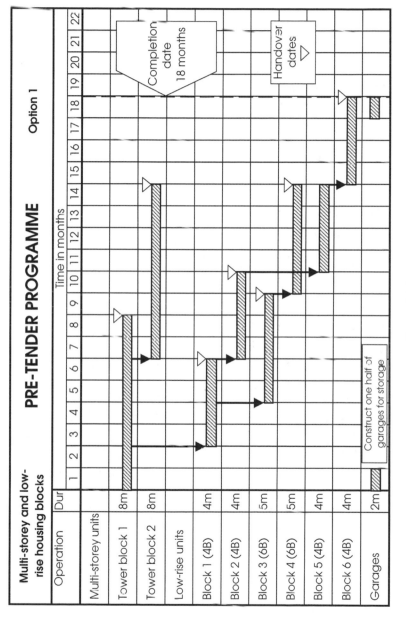

Figure 10.29

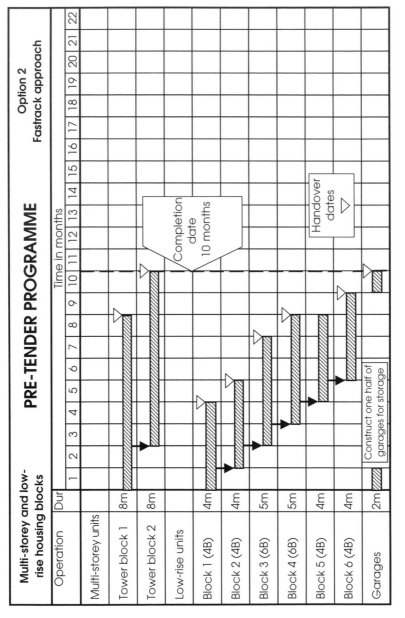

Figure 10.30

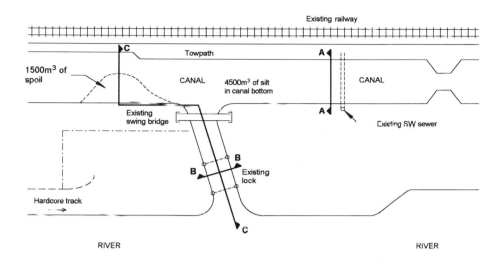

Canal Project General Arrangement (Existing)

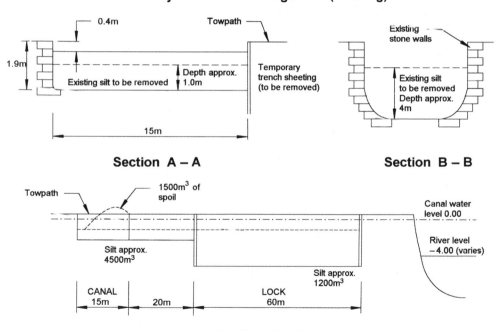

Section A – A

Section B – B

Section C – C

CANAL REFURBISHMENT PROJECT

Figure 10.31

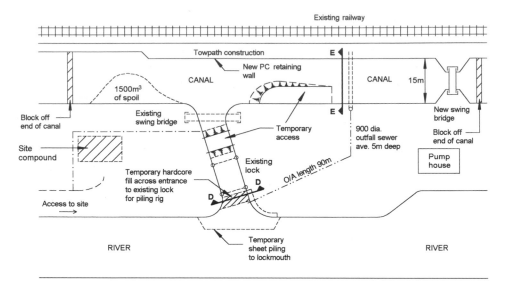

Canal Project General Arrangement (Proposed)

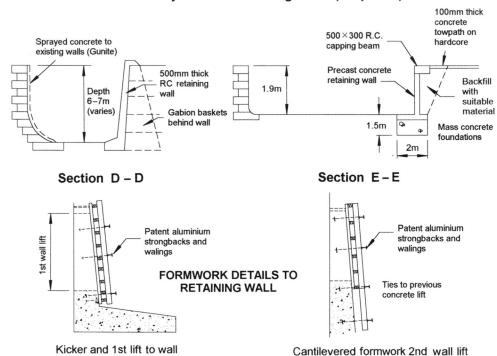

Section D – D

Section E – E

FORMWORK DETAILS TO RETAINING WALL

Kicker and 1st lift to wall

Cantilevered formwork 2nd wall lift

CANAL REFURBISHMENT PROJECT

Figure 10.32

The existing lock connects to an adjacent tidal river and is to be emptied of silt which is approximately 4 m deep. The silt is to be retained on site for landscape mounding. The canal basin is silted to a depth of 1 m which is to be removed to a tip off-site along with a spoil heap of 1500 m³ which is located at one end of the site.

A new in situ retaining wall is to be constructed to one side of the lock. The remaining walls are to be protected with a layer of sprayed concrete (gunite).

The existing canal is located adjacent to a live railway line and silt is to be cleaned out of the canal basin to a depth of approximately 1.5 m. This material is to be removed off-site to the contractor's tip.

A precast concrete retaining wall approximately 300 m long is to be constructed along the towpath next to the railway line. The wall is to be finished with an in situ concrete capping and a new concrete towpath laid alongside.

Worked examples

(1) Method statement

Access and temporary works are of primary importance to the development of a suitable construction sequence and safe system of working.

In order to carry out work in the existing canal and lock, the basins will have to be pumped dry and water excluded both from the remainder of the canal and the river. This will involve the construction of a sheet-piled cofferdam at the lockmouth, and both ends of the canal basin will have to be blocked or 'stanked-off' with sandbags or earth bunding.

The only access to the site is along an existing hardcore track. Temporary access is required to the canal basin and the pumphouse side of the lock and access for a piling rig is needed in order to construct a cofferdam to the lockmouth.

To do this, imported hardcore will be placed at the following locations:

- In the canal at the entrance to the lock to give access to the pumphouse side of the site
- In the main canal basin so as to form a ramp down to the bottom of the basin. This will be placed after cleaning out the silt and removed prior to removing the cofferdam
- In the lockmouth for the piling rig to stand on. Pollution of the river from the fill material may be a problem and the contractor would need to either consult the Environment Agency about this or consider driving the sheet piles from a pontoon located in the river

Figure 10.33 contains the contractor's method statement for the following operations:

METHOD STATEMENT

Operation	Quantity	Output	Duration	Method	Resources				Temp. works	Remarks
					Plant	Labour	S/C			
Construct cofferdam	350 m²	45 m² per day	Allow 2 weeks in total	Install sheet piles from temporary hardstanding using crawler crane and pile hammer. Pump out lock and fill to provide working area to install bracings. Remove hardstanding and install bottom walings and struts.	22RB BSP 900 Pile hammer Cat 325 excavator 2 No. 30T waggons	Piling Foreman Banksman 4 Labourers			Frodingham 3N sheet piles Steel I-Section bracing Quarry waste fill	
Clean out canal	600 m	30 m/hr	5 weeks	Excavate silt Load waggons and cart spoil to tip 8km from site. Provide dozer for levelling spoil at tip.	Komatsu 380 excavator 4 No. 30T waggons Cat D4 dozer	Ganger Banksman 2 Labourers				
Clean out lockmouth	1200 m of silt	12 m/hr	Say 3 weeks	Excavate silt from lockmouth. Load dumpers and cart to pumphouse area. Dozer to spread and level spoil in landscape mound.		Ganger Banksman 2 Labourers				

Figure 10.33a

METHOD STATEMENT

Operation	Quantity	Output	Duration	Method	Plant	Labour	S/C	Temp. works	Remarks
Drainage 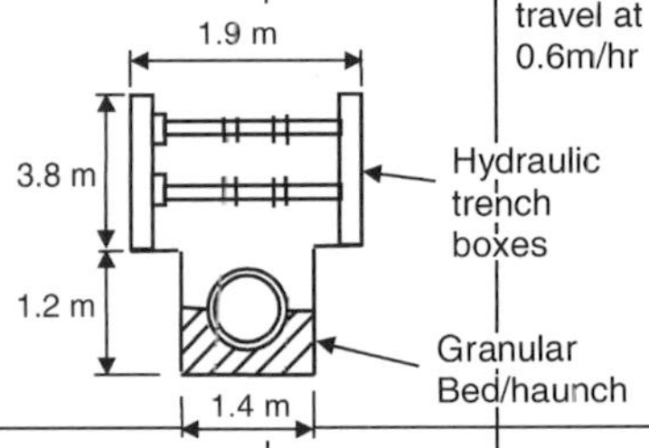	90 m 1.9 m 3.8 m 1.2 m 1.4 m Hydraulic trench boxes Granular Bed/haunch	Allow forward travel at 0.6m/hr	4 weeks	Excavate preliminary trench and install trench boxes. Excavate trench for pipe. Level bedding and lay and joint pipe. Backfill with suitable material and withdraw trench boxes in stages. Concrete bottom of M/H and lay bottom ring. Build up M/H and surround with concrete using prcprietory circular forms. Fix precast concrete 'bisquit' and heavy duty C.I. cover and frame.	Cat 325 Moxy dumper JCB 3CX Trench roller 4" pump Poker vibrator	Ganger 4 labourers Bricklayers		4 No. Hydraulic trench boxes 1 No. Hydraulic M/H box	900mm dia. sewer with 1No manhole approx depth 5m. Outfall at lockmouth
Precast retaining wall				Work sequence to be in 30m long bays ccmprising 10 No. precast concrete units per bay. Remove temporary trench sheeting and batter back excavation. Excavate foundations with back acter and pump mass concrete in foundation. Bed precast concrete units on foundation and backfill behind wall.	Mobile crane Poclain 75 excavator Poker vibrator Roller Concrete pump 30T waggon	Foreman Joiner Bricklayer Banksman 2 Labourers			
In situ retaining wall			Allow 5 weeks	Break out existing stone walls to lockmouth. Excavate for wall with battered dig. Fill and place gab on baskets behind wall. Shutter and fix, rebar in foundation, set ties in kicker and concrete. Fix rebar and shutter 1st wall lift with RMD formwork and ties, concrete and strip. Repeat sequence for 2nd lift.	Mobile crane Concrete pump Komatsu 380 Hydraulic breaker	Foreman 2 Joiners 2 Steel fixers 3 Labourers			

Figure 10.33b

- Cofferdam
- Clean out canal
- Clean out lockmouth
- Drainage
- Precast retaining wall
- In situ retaining wall

(2) Clause 14 programme

The ICE Conditions require:

- A programme for acceptance by the engineer showing the contractor's proposed order of construction (Clause 14 programme)
- A general description of arrangements and methods of construction
- Further detailed information from time to time regarding proposed methods of construction, temporary works and plant

The clause 14 programme has to be submitted for the engineer's consideration within 21 days of the award of the contract, together with details of the contractor's proposed method of working. This is shown in linked bar chart format in Figure 10.34, which illustrates the contractor's approach to the overall planning of the project.

(3) Risk assessment

The contractor's risk assessment for the construction of the sheet-piled cofferdam is shown in Figure 10.35.

(4) Safety method statement

The construction phase health and safety plan developed by the contractor should contain safety method statements for key activities on the contractor's programme, or those which pose potentially significant risks.

There are various formats that the contractor could use but a suggested approach for the construction of the sheet-piled cofferdam is shown in Figure 10.36. This has been based on the contractor's initial risk assessment for this activity, as illustrated in Figure 10.35.

10.7 Case study No. 5 – Highway project

General description

Figure 10.37 shows a proposed dual carriageway trunk road which is to be built to motorway standards under the Department of Transport Specification for Highway Works.

CANAL REFURBISHMENT

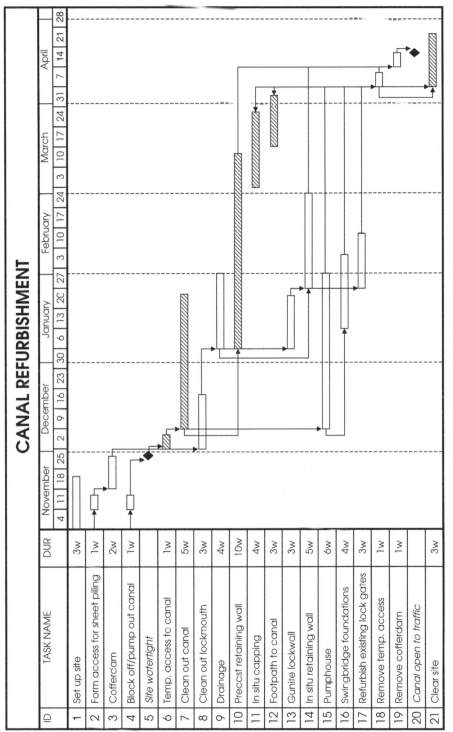

ID	TASK NAME	DUR
1	Set up site	3w
2	Form access for sheet piling	1w
3	Cofferdam	2w
4	Block off/pump out canal	1w
5	*Site watertight*	
6	Temp. access to canal	1w
7	Clean out canal	5w
8	Clean out lockmouth	3w
9	Drainage	4w
10	Precast retaining wall	10w
11	In situ capping	4w
12	Footpath to canal	3w
13	Gunite lockwall	3w
14	In situ retaining wall	5w
15	Pumphouse	6w
16	Swingbridge foundations	4w
17	Refurbish existing lock gates	3w
18	Remove temp. access	1w
19	Remove cofferdam	1w
20	*Canal open to traffic*	
21	Clear site	3w

CLAUSE 14 PROGRAMME

Figure 10.34

CONTRACTOR'S RISK ASSESSMENT

Contract		Contract No	Prepared by	Date of assessment
Canal renovation		C2314	GHM	23 May

Operation	Potential hazards	Risk assessment			Control measures
		H	M	L	
Construct sheet piled cofferdam	Trapping and crushing by moving plant		M		Warning device/banksman
	Work at height	H			Guard rails along lock
	Falls of materials	H			Leave piles proud for edge protection
	Confined space working			L	
	Working over water	H			Safety harness and life jacket to be worn

Programmed for	Training/certification	Action	PPE Required (specify)
16 June	All operatives to be CSCS certified	Method statement ☑	Safety harnesses
		Work permit ☑	Hard hats
	Plant operators to be CTA certified	Assessment :	Gloves
		COSHH ☐	Welding goggles
		Noise ☑	HV vests
			Life jackets
			Ear defenders

Figure 10.35

SAFETY METHOD STATEMENT

Contract	Contract No.	Prepared By	Date	Checked By	Date
Canal renovation	C2314	GHM	26 May	APH	27 May

Operation	Plant
Construct sheet piled cofferdam	22 RB Crane BSP 900 Pile hammer Komatsu 380 Excavator 2 No. 30T Wagons 150mm Diesel pump

Work sequence	Supervision and monitoring
Construct hardstanding for piling rig in lock mouth using imported quarry waste Erect guide frame and install Frodingham 3N piles Pump out water between hardstanding and piles Fill behind piles for access	Site engineer Piling foreman Operation to be monitored daily by the site agent Banksman to work with mobile plant Daily check on crane equipment, load indicators and operation

Work sequence	Controls
Remove piling frame Construct top bracing with steel wallings and struts Remove hardstanding and install bottom bracing Erect secure ladder access to bottom of cofferdam	Authorised personnel area only When working over water to remove piling frame, life jackets and safety harness must be worn Area to be fenced off at night with chestnut paled fence Warning notices to be displayed either side of cofferdam 'Danger deep excavation' and 'Danger deep water'

Emergency procedures	First Aid	PPE schedules
Send for site first aider and/or call emergency services where necessary Rescuers must not put themselves in danger Follow first aid drill it appropriate Do not remove evidence Notify site agent	Dinghy to be moored adjacent to cofferdam 2 No. Lifebuoys in wooden locker First aid box in site office 2-way radios	Safety harnesses Hard hats Gloves Welding goggles High visibility vests Life jackets Ear defenders

Figure 10.36

Under the ICE Conditions of Contract, the contractor is required to submit a clause 14 programme to the engineer within 21 days of the award of the contract. Additionally, if requested by the engineer, the contractor is also required to submit details of his proposed methods of construction, including temporary works and contractor's equipment. A time-chainage diagram would be an appropriate display on this type of project.

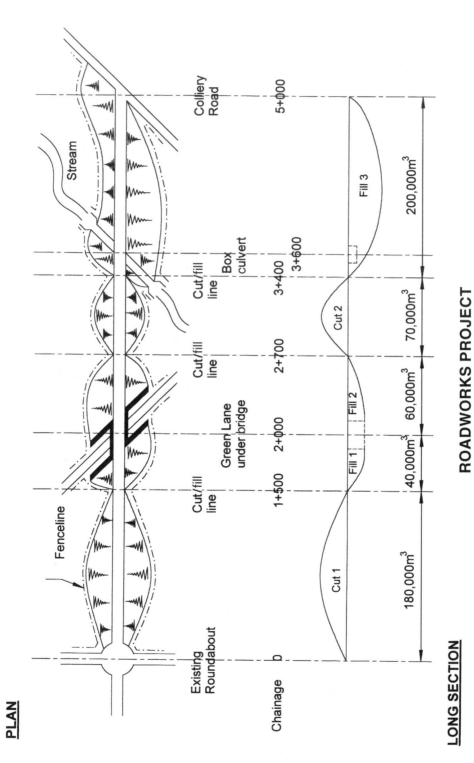

PLAN

Fenceline

Stream

Existing Roundabout

Cut/fill line

Green Lane under bridge

Cut/fill line

Cut/fill line

Box culvert

Colliery Road

Chainage 0 1+500 2+000 2+700 3+400 3+600 5+000

LONG SECTION

Cut 1 Fill 1 Fill 2 Cut 2 Fill 3

180,000m³ 40,000m³ 60,000m³ 70,000m³ 200,000m³

ROADWORKS PROJECT

Figure 10.37

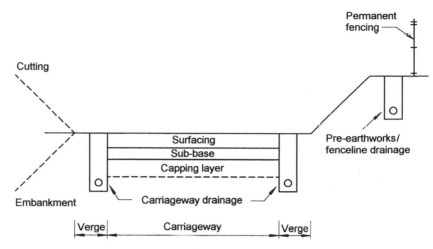

TYPICAL CROSS-SECTION

Figure 10.38

A typical cross-section through the road is given in Figure 10.38. This shows the permanent fencing and fenceline drainage which both have to be completed in advance of bulk earthworks operations. At chainage 3+600 there is a proposed box culvert which would also have to be completed in advance of filling to the embankment.

The main activities for the project and estimated durations are given in Figure 10.39.

There are three access points available on to the site, one at the existing roundabout, one at Colliery Road and the other via Green Lane.

Worked examples

(1) Construction strategy

Earthworks
The project is significantly influenced by the earthworks in that the sequencing of cut and fill operations largely controls the sequencing of the entire project. However, no activity can be considered mutually exclusive as the sequencing of the earthworks is itself affected by the construction of the bridge and box culvert.

On major road projects, the quantities of muck to be shifted are often very large and two or three million cubic metres is not unusual. The quantities for this case study are therefore relatively small but there are a number of problems which face the earthworks contractor:

- The major cut and fill zones are at opposite ends of the site and the average haul distance on the project is approximately 2 km each way

ACTIVITY	DURATION (weeks)
Set up site	6
Fencing and site clearance	10
Topsoil strip	4
Fenceline drainage	14
Excavation and filling	20
Capping layer	11
Carriageway drainage	13
Sub-base	11
Surfacing	13
Box culvert	8
Bridge	26
Finishes	19
Tie-ins to existing roads	6

SCHEDULE OF ACTIVITIES

Figure 10.39

- The second fill zone is bisected by a proposed bridge site
- A box culvert is required at chainage 3+600 which must precede filling to the embankment
- There is a shortfall of acceptable fill material and therefore 50 000 m^3 will have to be imported
- The existing road at Green Lane must be kept open to traffic during construction of the new bridge

Earthworks plant for a project of this nature has changed over the last 30 years or so. Conventionally, major earthworks were carried out using rope-operated scraper boxes towed by caterpillar-type tractors, but motorised scrapers, which were faster and with larger capacity, became popular. Scrapers often need dozers pushing to help loading, especially in heavy ground conditions.

Nowadays, it is also popular to use large tracked back-actors with articulated dump trucks (ADTs) carrying excavated materials to fill zones. Modern excavators are fast and powerful and the new breed of dumpers carry large volumes of spoil. However, where the haul distances are long, this either means the excavators being kept waiting or employing a large fleet of dumpers. A contractor may well use a combination of scrapers and excavators/dumpers, the main considerations being:

- Volume of earthmoving
- Nature of materials to be removed
- Whether material is to be used for fill or carted off site
- Cost of delivering plant to site
- Cost and availability of plant
- Use of own or hired plant
- Haul distances
- Time constraints
- Access
- Restrictions (e.g. bridges)

Drainage
The drainage works comprise:

- Pre-earthworks or fenceline drainage – mainly consisting of porous rubble-filled 'french' drains with associated catchpits and connections to surface water ('carrier') drains, culverts and lined and unlined ditches
- Carriageway drainage – comprising both 'carrier' drains, manholes and gullies and 'french' drains and catchpits

The pre-earthworks drainage normally commences early in the project and follows on behind erection of the permanent fenceline and the topsoil strip over the site.

When the pre-earthworks drainage is reasonably advanced, bulk earthworks can follow on behind. Earthworks may be interrupted by culverts crossing the carriageway, which will affect muck-shift sequencing and may involve temporary diversions around the site of the culvert.

Pre-earthworks drainage poses problems for the contractor such as:

- **Access** – There may be no haul roads in and plant may have to 'track' along from the main access points, e.g. access for the box culvert would have to be from Colliery Road
- **Materials** – Without proper access for road vehicles, site conditions and the weather will very much dictate whether access will be possible along the spread of the job. Drainage materials such as pipes, manhole/catchpit components, bricks and ironwork may have to be delivered to the main compound and then doublehandled by site transport such as tractor and trailer to the various locations along the job. Drain fills are delivered from the quarry by road transport and may have to be tipped at access points and reloaded on to site dumpers. Ready-mixed concrete may similarly be a problem
- **Drainage arisings** – A major proportion of the spoil from drainage excavations will not be required for filling trenches but may not be

suitable for filling elsewhere on the site. The suitability of fills is the responsibility of the main contractor/earthworks subcontractor and therefore the drainage subcontractor will have to be instructed as to what is required of him. Part of the contractor's thinking could be either for the drainage subcontractor to load the arisings into waggons/dumpers or, perhaps more cost effectively, for the arisings to be left to one side to be removed by the earthworks sub-contractor. This is an interface which would require managing by the main contractor, and the respective responsibilities would have to be written into each of the subcontract packages

- **Resources** – The main contractor will have to carefully manage the drainage subcontractor to make sure sufficient gangs are available so as not to slow the earthworks down. This is important because the earthworks 'season' is relatively short. Often on large projects the drainage is split into two or more packages and different sub-contractors are employed on each. Again there will be interface problems here, particularly with regard to the control of materials, which the main contractor has to think about

Structures

One of the problems facing the contractor is the construction of struc-tures across the carriageway, such as bridges and culverts, especially in fill areas. In this example there is a bridge at chainage 2+000 and a box culvert at chainage 3+600, both of which would have to be constructed in advance of embankment construction. Fill can then be brought up to and behind each structure so as to bring levels up to the correct height for road construction to follow.

With regard to the bridge at chainage 2+000, abutment walls and foundations will have to be constructed parallel to Green Lane. This may require traffic management arrangements or possibly a tempor-ary diversion of traffic around the bridge site. Alternatively, it may be possible to close the road during construction.

In order that bridge construction can proceed without impeding earthworks operations, a temporary haul road will be constructed around the bridge site. This will involve traffic control measures on Green Lane as heavy plant will be crossing frequently. This will also complicate the time-chainage programme as drainage works and other activities will have to be left incomplete at this location until later. This has not been shown on Figure 10.40 for simplicity.

Capping layer

Depending on the design of the roadworks, a layer of fill may be required as a foundation for road construction. This is known as the capping layer and this operation follows on from bulk earthworks.

Due to the earthworks sequencing on this particular project, the capping layer, drainage and road construction would start at chainage

TIME-CHAINAGE DIAGRAM

Roadworks Project

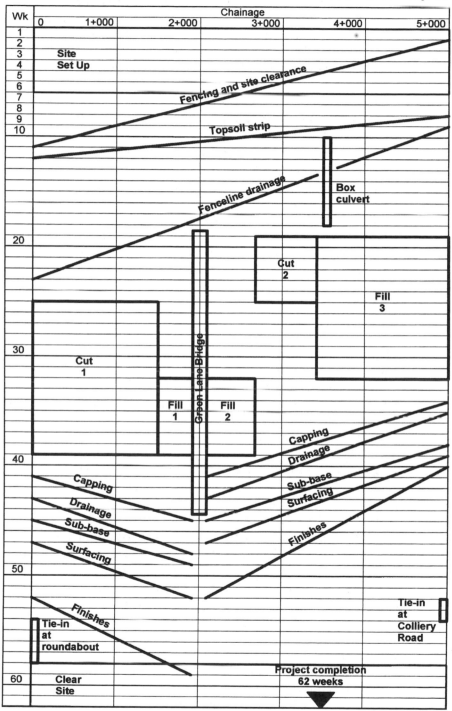

NB: Holidays and other activities omitted for clarity

Figure 10.40

5+000 and progress towards chainage 2+000. This would be more efficient than starting from chainage 0+000 as capping etc. could follow filling to zone 3.

(2) Method statement (simplified)

Earthworks
Bulk excavation to Cut 1 assuming all material is acceptable as fill. Part of Cut 1 is used for embankment construction at Fill 3 and the remainder for bringing up embankment construction either side of Green Lane Bridge.

Quantity = 180 000 m^3 of excavation

Method – Cutting to be taken out using a combination of motor scrapers push-loaded with a tracked dozer and 40 tonne 360° tracked excavators loading into articulated dump trucks.

Average haul distance = approx 2.25 km.

Filling and compaction is a separate activity.

Outputs – The average haul distance is relatively long and the scrapers will do well to achieve three cycles per hour. The scraper crews usually work a long day in the order of 10 working hours over 6 or 7 days per week in the earthworks season. A large scraper will move an average of 450 m^3 each per day over the haul distances on this project.

The output of the back actors is usually governed by the haul times of the dumpers, unless the digging is hard. Therefore the output is decided on how many trips/cycles can be achieved. Again, because of the average haul distance, this will be no more than three per hour giving around 300 m^3 per dumper per day. Using three dumpers this equates to 90 m^3 per hour for the excavators, which is no problem for two large machines. The gang therefore has capacity for trimming operations and dealing with obstructions, etc.

Gang:

Foreman
3 No. Terex TS24 motor scrapers
1 No. Cat D7 dozer (pushing)
1 No. Akerman H10 excavator
1 No. Komatsu PC380 excavator
3 No. Cat D400 dumpers
Banksman

Activity duration:

$$= \frac{180\,000\ \text{m}^3}{[(3\ \text{scrapers} \times 450\ \text{m}^3) + (3\ \text{dumpers} \times 300\ \text{m}^3)] \times 6\ \text{days}}$$

$= \textbf{14 weeks}$

Drainage

Assume the carriageway drainage from chainage 0+000 to 2+000 varies from 150–300 mm in diameter at an average depth of 1.50 m with 1 No. catchpit at approximately 100 m intervals

Quantity = 4500 m of pipework including crossings and
connections
40 No. catchpits

Method:

1 No. drainlaying gang either side of the carriageway excavating trenches, laying and jointing pipes and backfilling.

1 No. catchpit gang following up concreting catchpit bases and placing bottom chamber ring and then next day completing remaining rings, precast slab and temporary cover.

Gullies, gully connections and brickwork and ironwork to gullies and catchpits to follow on later during roadworks.

Gangs:

Drainlaying (2 gangs):
        Ganger
        Komatsu PC180 tracked excavator
        Banksman
        Komatsu WA470 wheeled loader
        3 labourers
Catchpits: Ganger
        JCB 3CX
        Banksman
        Bricklayer
        2 labourers

Outputs:

Drainlaying: 90 m/gang/day working a 5-day week
Catchpits: Average 2 No. per day

Activity duration:

$$\text{Drainlaying} = \frac{4500 \text{ m}}{90 \text{ m/day} \times 5 \text{ days} \times 2 \text{ gangs}} = \textbf{5 weeks}$$

$$\text{Catchpits} = \frac{40 \text{ No.}}{\text{Av. 2 No. per day} \times 5 \text{ days}} = \textbf{4 weeks}$$

The two operations run concurrently and therefore the duration for the drainage activity is 5 weeks overall.

Structures
3000 mm × 2500 mm precast concrete box culvert at chainage 3+600 comprising 1200 mm long units laid in trench on gravel bed jointed with compressible neoprene and polysulphide sealant and backfilled with free draining material.

> Quantity = 80 linear metres
>   1 No. headwall at each end

Method – Excavate trench with battered sides using tracked excavator and remove material to spoil with articulated dump trucks. Lay bed of gravel and lay and joint precast concrete units to line and level. Backfill working space on completion. Headwalls to be constructed at each end of the culvert comprising gabion baskets filled with stone. The excavator will be used for diverting the stream and for all excavation, backfilling and handling bedding materials. A rough terrain mobile crane will be needed for offloading and handling the precast concrete culvert sections and for lifting and placing the gabions.

Gang:

Foreman
Cat 320 tracked excavator
40T mobile telescopic crane
Banksman
Volvo A25 dumper
3 labourers

Output:

Allow forward travel of three complete precast sectional units per day
    Activity duration:

> Diversion of stream   = 1 week

$$\text{Culverting} = \frac{80 \text{ m}}{3 \text{ units/day} \times 5 \text{ days}} = 5 \text{ weeks}$$

Headwalls	= 2 weeks
Total	**= 8 weeks**

Capping layer
Acceptable capping layer material will be imported from a quarry located 10 miles from the site. It is assumed that the necessary quantities of material will be available to suit site production outputs

Quantity = 12 000 m^3 of fill (chainage 0 – 1 + 500)

Method – Imported material to be delivered to site in 20 tonne road waggons via the existing roundabout at chainage 0+000. The material is to be tipped and dozed into position with a dozer blade and spread in 200 mm layers with a motor grader. Compaction is to be carried out using a sit-on tandem vibrating roller. Final trim to be carried out immediately prior to laying sub-base material.

Outputs – Haulage distance and availability of waggons will largely dictate site output and the contractor will have to consider the possibility of more than one supplier for such large quantities. The gang is expected to place 300 m^3 in a 10-hour working day under normal conditions.

Gang:

Ganger
Cat D6N dozer
Cat 120H motor grader
Bomag BW125 roller
Banksman

$$\text{Activity duration} = \frac{12\ 000\ \text{m}^3}{600\ \text{m}^3/\text{day} \times 5\ \text{days}} = \textbf{4 weeks}$$

(3) Clause 14 programme

Earthworks
The earthworks activity will occupy Cut 1 for 14 weeks between chainages 0+000 and 1+500 and this is shown plotted on the time-chainage diagram in Figure 10.40.

Drainage
Calculating the slope of linear activities requires consideration of:

• Quantity of work
• Method of working

- Gang size/make-up
- Estimated gang output (rate of forward travel)

The drainage activity is plotted between chainage 0+000 commencing at week 43 and chainage 2+000 finishing at week 48 (see Figure 10.40).

Structures
The culvert is a 'static' activity and is therefore plotted at chainage 3+600 with a duration of 8 weeks, which is shown on the vertical axis of the time-chainage diagram (Figure 10.40).

Capping layer
A 2-week lead time has been allowed after completion of filling to zone 3 in order to give a time buffer between filling and capping to zone 2 at chainage 2+600.

At week 41, capping layer and drainage etc. are programmed to commence at chainage 0+000, working towards Green Lane bridge. Note that continuity of working is preserved by using time buffers of different durations between activities (see Figure 10.40).

Reference

Walker, A. (2002) *Project Management in Construction*, 4th edn. Blackwell Publishing.

Index